Universitext

*Editorial Board
(North America)*

S. Axler
F.W. Gehring
K.A. Ribet

Springer
*New York
Berlin
Heidelberg
Barcelona
Hong Kong
London
Milan
Paris
Singapore
Tokyo*

Universitext

Editors (North America): S. Axler, F.W. Gehring, and K.A. Ribet

Aksoy/Khamsi: Nonstandard Methods in Fixed Point Theory
Andersson: Topics in Complex Analysis
Aupetit: A Primer on Spectral Theory
Berberian: Fundamentals of Real Analysis
Booss/Bleecker: Topology and Analysis
Borkar: Probability Theory: An Advanced Course
Böttcher/Silbermann: Introduction to Large Truncated Toeplitz Matrices
Carleson/Gamelin: Complex Dynamics
Cecil: Lie Sphere Geometry: With Applications to Submanifolds
Chae: Lebesgue Integration (2nd ed.)
Charlap: Bieberbach Groups and Flat Manifolds
Chern: Complex Manifolds Without Potential Theory
Cohn: A Classical Invitation to Algebraic Numbers and Class Fields
Curtis: Abstract Linear Algebra
Curtis: Matrix Groups
DiBenedetto: Degenerate Parabolic Equations
Dimca: Singularities and Topology of Hypersurfaces
Edwards: A Formal Background to Mathematics I a/b
Edwards: A Formal Background to Mathematics II a/b
Foulds: Graph Theory Applications
Friedman: Algebraic Surfaces and Holomorphic Vector Bundles
Fuhrmann: A Polynomial Approach to Linear Algebra
Gardiner: A First Course in Group Theory
Gårding/Tambour: Algebra for Computer Science
Goldblatt: Orthogonality and Spacetime Geometry
Gustafson/Rao: Numerical Range: The Field of Values of Linear Operators and Matrices
Hahn: Quadratic Algebras, Clifford Algebras, and Arithmetic Witt Groups
Holmgren: A First Course in Discrete Dynamical Systems
Howe/Tan: Non-Abelian Harmonic Analysis: Applications of $SL(2, R)$
Howes: Modern Analysis and Topology
Hsieh/Sibuya: Basic Theory of Ordinary Differential Equations
Humi/Miller: Second Course in Ordinary Differential Equations
Hurwitz/Kritikos: Lectures on Number Theory
Jennings: Modern Geometry with Applications
Jones/Morris/Pearson: Abstract Algebra and Famous Impossibilities
Kannan/Krueger: Advanced Analysis
Kelly/Matthews: The Non-Euclidean Hyperbolic Plane
Kostrikin: Introduction to Algebra
Luecking/Rubel: Complex Analysis: A Functional Analysis Approach
MacLane/Moerdijk: Sheaves in Geometry and Logic
Marcus: Number Fields
McCarthy: Introduction to Arithmetical Functions
Meyer: Essential Mathematics for Applied Fields
Mines/Richman/Ruitenburg: A Course in Constructive Algebra
Moise: Introductory Problems Course in Analysis and Topology
Morris: Introduction to Game Theory
Polster: A Geometrical Picture Book
Porter/Woods: Extensions and Absolutes of Hausdorff Spaces

(continued after index)

Fuzhen Zhang

Matrix Theory

Basic Results and Techniques

 Springer

Fuzhen Zhang
Department of Math, Science, and Technology
Nova Southeastern University
Fort Lauderdale, FL 33314
USA

Editorial Board
(North America)

S. Axler
Mathematics Department
San Francisco State University
San Francisco, CA 94132
USA

F.W. Gehring
Mathematics Department
East Hall
University of Michigan
Ann Arbor, MI 48109
USA

K.A. Ribet
Department of Mathematics
University of California at Berkeley
Berkeley, CA 94720-3840
USA

Mathematics Subject Classification (1991): 15-01

Library of Congress Cataloging-in-Publication Data
Zhang, Fuzhen, 1961–
  Matrix theory : basic results and techniques / Fuzhen Zhang.
    p.   cm. — (Universitext)
  Includes bibliographical references and index.
  ISBN 0-387-98696-0 (hc : alk. paper)
  1. Matrices.   I. Title
  QA188.Z47   1999
  512.9´434—dc21   98-51754

Printed on acid-free paper.

© 1999 Springer-Verlag New York Inc.
All rights reserved. This work may not be translated or copied in whole or in part without the written permission of the publisher (Springer-Verlag New York, Inc., 175 Fifth Avenue, New York, NY 10010, USA), except for brief excerpts in connection with reviews or scholarly analysis. Use in connection with any form of information storage and retrieval, electronic adaptation, computer software, or by similar or dissimilar methodology now known or hereafter developed is forbidden.
The use of general descriptive names, trade names, trademarks, etc., in this publication, even if the former are not especially identified, is not to be taken as a sign that such names, as understood by the Trade Marks and Merchandise Marks Act, may accordingly be used freely by anyone.

Production managed by Terry Kornak; manufacturing supervised by Jeffrey Taub.
Camera-ready copy prepared by the author.
Printed and bound by R. R. Donnelley and Sons, Harrisonburg, VA.
Printed in the United States of America.

9 8 7 6 5 4 3 2 1

ISBN 0-387-98696-0   Springer-Verlag New York Berlin Heidelberg   SPIN 10707272

To my wife Cheng, daughter Sunny, and son Andrew

Preface

It has been my goal to write a concise book that contains fundamental ideas, results, and techniques in linear algebra and (mainly) in matrix theory which are accessible to general readers with an elementary linear algebra background. I hope this book serves the purpose.

Having been studied for more than a century, linear algebra is of central importance to all fields of mathematics. Matrix theory is widely used in a variety of areas including applied math, computer science, economics, engineering, operations research, statistics, and others.

Modern work in matrix theory is not confined to either linear or algebraic techniques. The subject has a great deal of interactions with combinatorics, group theory, graph theory, operator theory, and other mathematical disciplines. Matrix theory is still one of the richest branches of mathematics; some intriguing problems in the field were long standing, such as the Van der Warden conjecture (1926–1980), and some, such as the permanental-dominance conjecture (since 1966), are still open.

This book contains eight chapters covering various topics from similarity and special types of matrices to Schur complements and matrix normality. Each chapter focuses on the results, techniques, and methods that are beautiful, interesting, and representative, followed by carefully selected problems. Many theorems are given different proofs. The material is treated primarily by matrix approaches and reflects the author's tastes.

The book can be used as a text or a supplement for a linear algebra or matrix theory class or seminar. A one-semester course may consist of the first four chapters plus any other chapter(s) or section(s). The only prerequisites are a decent background in elementary linear algebra and calculus (continuity, derivative, and compactness in a few places). The book can also serve as a reference for researchers and instructors.

The author has benefited from numerous books and journals, including *The American Mathematical Monthly, Linear Algebra and Its Applications, Linear and Multilinear Algebra,* and the International Linear Algebra Society (ILAS) Bulletin *Image*. This book would not exist without the earlier works of a great number of authors (see the References).

I am grateful to the following professors for many valuable suggestions and input and for carefully reading the manuscript so that many errors have been eliminated from the earlier version of the book:

Professor R. B. Bapat (Indian Statistical Institute),
Professor L. Elsner (University of Bielefeld),
Professor R. A. Horn (University of Utah),

Professor T.-G. Lei (National Natural Science Foundation of China),
Professor J.-S. Li (University of Science and Technology of China),
Professor R.-C. Li (University of Kentucky),
Professor Z.-S. Li (Georgia State University),
Professor D. Simon (Nova Southeastern University),
Professor G. P. H. Styan (McGill University),
Professor B.-Y. Wang (Beijing Normal University), and
Professor X.-P. Zhang (Beijing Normal University).

F.-Z. Zhang
Ft. Lauderdale
March 5, 1999
zhang@polaris.nova.edu
http://www.polaris.nova.edu/~zhang

Contents

Preface vii
Frequently Used Notation and Terminology xi
Frequently Used Theorems xiii

1 Elementary Linear Algebra Review   1

    1.1   Vector Spaces 1
    1.2   Matrices 6
    1.3   Linear Transformations and Eigenvalues 14
    1.4   Inner Product Spaces 22

2 Partitioned Matrices   29

    2.1   Elementary Operations of Partitioned Matrices 29
    2.2   The Determinant and Inverse of Partitioned Matrices 36
    2.3   The Inverse of a Sum 43
    2.4   The Rank of Product and Sum 46
    2.5   Eigenvalues of AB and BA 51
    2.6   The Continuity Argument 56

3 Matrix Polynomials and Canonical Forms   59

    3.1   Commuting Matrices 59
    3.2   Matrix Decompositions 64
    3.3   Annihilating Polynomials of Matrices 70
    3.4   Jordan Canonical Forms 74
    3.5   The Matrices A^T, $\overline{A}$, A^*, A^TA, A^*A, and $\overline{A}A$ 83
    3.6   Numerical Range 88

4 Special Types of Matrices   93

    4.1   Idempotence, Nilpotence, Involution, and Projection 93
    4.2   Tridiagonal Matrices 101
    4.3   Circulant Matrices 106
    4.4   Vandermonde Matrices 111
    4.5   Hadamard Matrices 118
    4.6   Permutation and Doubly Stochastic Matrices 123

5 Unitary Matrices and Contractions — 131

- 5.1 Properties of Unitary Matrices 131
- 5.2 Real Orthogonal Matrices 137
- 5.3 Metric Space and Contractions 142
- 5.4 Contractions and Unitary Matrices 148
- 5.5 The Unitary Similarity of Real Matrices 152
- 5.6 A Trace Inequality of Unitary Matrices 155

6 Positive Semidefinite Matrices — 159

- 6.1 Positive Semidefinite Matrices 159
- 6.2 A Pair of Positive Semidefinite Matrices 166
- 6.3 Partitioned Positive Semidefinite Matrices 175
- 6.4 Schur Complements and Determinantal Inequalities 184
- 6.5 The Kronecker Product and Hadamard Product 190
- 6.6 Schur Complements and Hadamard Products 198
- 6.7 The Cauchy-Schwarz and Kantorovich Inequalities 203

7 Hermitian Matrices — 208

- 7.1 Hermitian Matrices 208
- 7.2 The Product of Hermitian Matrices 213
- 7.3 The Min-Max Theorem and Interlacing Theorem 219
- 7.4 Eigenvalue and Singular Value Inequalities 227
- 7.5 A Triangle Inequality for the Matrix $(A^*A)^{\frac{1}{2}}$ 235

8 Normal Matrices — 240

- 8.1 Equivalent Conditions 240
- 8.2 Normal Matrices with Zero and One Entries 251
- 8.3 A Cauchy-Schwarz Type Inequality for Matrix $(A^*A)^{\frac{1}{2}}$ ... 255
- 8.4 Majorization and Matrix Normality 260

References 265
Notation 273
Index 275

Frequently Used Notation and Terminology

$\mathbb{M}_n$	$n \times n$ matrices with complex number entries
$\mathbb{C}^n$	column vectors of n complex components
$\dim V$	dimension of vector space V
(u, v)	inner product of vectors u and v
$\|x\|$	norm or length of vector x, i.e., $\|x\| = \left(\sum_i |x_i|^2\right)^{\frac{1}{2}}$
I	identity matrix
$A = (a_{ij})$	matrix A with entries a_{ij}
rank (A)	rank of matrix A
tr A	trace of matrix A
det A	determinant of matrix A
A^{-1}	inverse of matrix A
A^T	transpose of matrix A
$\overline{A}$	conjugate of matrix A
A^*	conjugate transpose of matrix A, i.e., $A^* = \overline{A}^T$
$|A|$	determinant for a block matrix A or matrix $(A^*A)^{\frac{1}{2}}$
$[A]$	principal submatrix of matrix A
Ker(A)	kernel or null of A, i.e., Ker$(A) = \{x : Ax = 0\}$
Im(A)	image space of A, i.e., Im$(A) = \{Ax\}$
$\lambda_{\max}(A)$	largest eigenvalue of matrix A
$\sigma_{\max}(A)$	largest singular value of matrix A
$A \geq 0$	A is positive semidefinite
$A \geq B$	$A - B$ is positive semidefinite
diag$(\lambda_1, \lambda_2, \ldots, \lambda_n)$	diagonal matrix with $\lambda_1, \lambda_2, \ldots, \lambda_n$ on the diagonal
$A \circ B$	Hadamard product of matrices A and B
$A \otimes B$	Kronecker product of matrices A and B

An $n \times n$ matrix A is said to be

upper-triangular	if all entries below the diagonal are zero
diagonalizable	if $P^{-1}AP$ is diagonal for some invertible matrix P
similar to B	if $P^{-1}AP = B$ for some invertible matrix P
unitarily similar to B	if $U^*AU = B$ for some unitary matrix U
unitary	if $AA^* = A^*A = I$
positive semidefinite	if $x^*Ax \geq 0$ for all vectors $x \in \mathbb{C}^n$
Hermitian	if $A = A^*$
normal	if $A^*A = AA^*$

$\lambda \in \mathbb{C}$ is an eigenvalue of $A \in \mathbb{M}_n$ if $Ax = \lambda x$ for some nonzero $x \in \mathbb{C}^n$.

Frequently Used Theorems

- **Cauchy-Schwarz inequality:** Let V be an inner product space over a number field ($\mathbb{R}$ or $\mathbb{C}$). Then for all x and y in V

$$|(x,y)|^2 \leq (x,x)(y,y).$$

  Equality holds if and only if x and y are linearly dependent.

- **Theorem on the eigenvalues of AB and BA:** Let A and B be $m \times n$ and $n \times m$ complex matrices, respectively. Then AB and BA have the same nonzero eigenvalues, counting multiplicity. Thus

$$\operatorname{tr}(AB) = \operatorname{tr}(BA).$$

- **Schur triangularization theorem:** For any square matrix A, there exists a unitary matrix U such that U^*AU is upper-triangular.

- **Jordan decomposition theorem:** Let A be an n-square complex matrix. Then there exists an $n \times n$ invertible matrix P such that

$$A = P^*(J_1 \oplus J_2 \oplus \cdots \oplus J_k)P,$$

  where each J_i, $i = 1, 2, \ldots, k$, is a Jordan block.

- **Spectral decomposition theorem:** Let A be an $n \times n$ normal matrix with eigenvalues $\lambda_1, \lambda_2, \ldots, \lambda_n$. Then there exists an $n \times n$ unitary matrix U such that

$$A = U^* \operatorname{diag}(\lambda_1, \lambda_2, \ldots, \lambda_n) U.$$

  In particular, if A is positive semidefinite, then $\lambda_i \geq 0$; if A is Hermitian, then the λ_i are real; and if A is unitary, then $|\lambda_i| = 1$.

- **Singular value decomposition theorem:** Let A be an $m \times n$ complex matrix with rank r. Then there exist an $m \times m$ unitary matrix U and an $n \times n$ unitary matrix V such that

$$A = UDV,$$

  where D is the $m \times n$ matrix with (i,i)-entries the singular values of A, $i = 1, 2, \ldots, r$, and other entries 0. If $m = n$, then D is diagonal.

CHAPTER 1

Elementary Linear Algebra Review

INTRODUCTION We briefly review, mostly without proof, the basic concepts and results taught in an elementary linear algebra course. The subjects are vector spaces, basis and dimension, linear transformations, the similarity of matrices, and inner product spaces.

1.1 Vector Spaces

Let V be a set of objects and let $\mathbb{F}$ be a field, mostly the real number field $\mathbb{R}$ or the complex number field $\mathbb{C}$ throughout this book. The set V is called a *vector space* over $\mathbb{F}$ if the operations *addition*

$$u + v, \quad u, v \in V,$$

and *scalar multiplication*

$$cv, \quad c \in \mathbb{F}, v \in V,$$

are defined so that the addition is associative, is commutative, has an additive identity 0 and additive inverse $-v$ in V for each $v \in V$, and so that the scalar multiplication is distributive, is associative, and has an identity 1 for which $1v = v$ for every $v \in V$.

The elements of a vector space are called *vectors*.

For example, $\mathbb{R}^n$, the set of all real column vectors

$$\begin{pmatrix} x_1 \\ x_2 \\ \vdots \\ x_n \end{pmatrix}, \quad \text{written as} \quad (x_1, x_2, \ldots, x_n)^T,$$

is a vector space over $\mathbb{R}$ with respect to the addition

$$(x_1, x_2, \ldots, x_n)^T + (y_1, y_2, \ldots, y_n)^T = (x_1 + y_1, x_2 + y_2, \ldots, x_n + y_n)^T$$

and the scalar multiplication

$$c\,(x_1, x_2, \ldots, x_n)^T = (cx_1, cx_2, \ldots, cx_n)^T.$$

Note that the real row vectors also form a vector space over $\mathbb{R}$. For convenience, we reserve $\mathbb{F}^n$ for the column vectors in future discussions involving the matrix and vector product Ax.

Let $S = \{v_1, v_2, \ldots, v_k\}$ be a subset of a vector space V over a field $\mathbb{F}$. Denote by Span S the collection of all linear combinations of the vectors in S, that is,

$$\text{Span}\,S = \{c_1v_1 + c_2v_2 + \cdots + c_kv_k : \text{each } c_i \in \mathbb{F} \text{ and } v_i \in S\}.$$

The set Span S is also a vector space over $\mathbb{F}$. If Span $S = V$, then every vector in V can be expressed as a linear combination of vectors in S. In such cases we say that the set S *spans* the vector space V.

A set $S = \{v_1, v_2, \ldots, v_k\}$ is said to be *linearly independent* if the vector equation

$$c_1v_1 + c_2v_2 + \cdots + c_kv_k = 0$$

has only the trivial solution $c_1 = c_2 = \cdots = c_k = 0$. If there are also nontrivial solutions, then S is said to be *linearly dependent*.

For instance, both $\{(1,0)^T, (0,1)^T, (1,1)^T\}$ and $\{(1,0)^T, (0,1)^T\}$ span $\mathbb{R}^2$. The first set is linearly dependent and the second set is linearly independent.

A *basis* of a vector space V is a linearly independent set that spans V. If V possesses a basis of an n-vector set $S = \{v_1, v_2, \ldots, v_n\}$, we say that V is of *dimension n*, written as $\dim V = n$. Conventionally, the dimension of a zero vector space is 0. If any finite set cannot span V, then V is infinite-dimensional.

For instance, $\mathbb{C}$ is a vector space of dimension 2 over $\mathbb{R}$ with basis $\{1, i\}$, where $i = \sqrt{-1}$, and of dimension 1 over $\mathbb{C}$ with basis $\{1\}$.

$\mathbb{C}^n$, the set of column vectors of n complex components, is a vector space over $\mathbb{C}$ having *standard basis*

$$e_1 = (1, 0, \ldots, 0, 0)^T,$$

$$e_2 = (0, 1, \ldots, 0, 0)^T,$$

$$\vdots$$

$$e_n = (0, 0, \ldots, 0, 1)^T.$$

If $\{u_1, u_2, \ldots, u_n\}$ is a basis for a vector space V of dimension n, then every x in V can be uniquely expressed as a linear combination of the basis vectors:

$$x = x_1 u_1 + x_2 u_2 + \cdots + x_n u_n,$$

where the x_i are scalars. The n-tuple $(x_1, x_2, \ldots, x_n)$ is called the *coordinate* of vector x with respect to the basis.

Let V be a vector space of dimension n, and let $\{v_1, v_2, \ldots, v_k\}$ be a linearly independent subset of V. Then $k \leq n$, and it is not difficult to see that if $k < n$, then there exists a vector $v_{k+1} \in V$ such that the set $\{v_1, v_2, \ldots, v_k, v_{k+1}\}$ is linearly independent (Problem 13). It follows that the set $\{v_1, v_2, \ldots, v_k\}$ can be extended to a basis of V.

Let W be a subset of a vector space V. If W is also a vector space under the addition and the scalar multiplication for V, then W is called a *subspace* of V.

For subspaces V_1 and V_2, the *sum* of V_1 and V_2 is defined to be

$$V_1 + V_2 = \{v_1 + v_2 : v_1 \in V_1, v_2 \in V_2\}.$$

It follows that the sum $V_1 + V_2$ is also a subspace. In addition, the intersection $V_1 \cap V_2$ is a subspace, and

$$V_1 \cap V_2 \subseteq V_i \subseteq V_1 + V_2, \quad i = 1, 2.$$

Theorem 1.1 (Dimension Identity) *Let V be a finite-dimensional vector space, and let V_1 and V_2 be subspaces of V. Then*

$$\dim V_1 + \dim V_2 = \dim(V_1 + V_2) + \dim(V_1 \cap V_2).$$

The proof is done by first choosing a basis $\{u_1, \ldots, u_k\}$ for $V_1 \cap V_2$, extending it to a basis $\{u_1, \ldots, u_k, v_{k+1}, \ldots, v_s\}$ for V_1 and a basis $\{u_1, \ldots, u_k, w_{k+1}, \ldots, w_t\}$ for V_2, and then showing that

$$\{u_1, \ldots, u_k, v_{k+1}, \ldots, v_s, w_{k+1}, \ldots, w_t\}$$

is a basis for $V_1 + V_2$.

It follows that subspaces V_1 and V_2 contain nonzero common vectors if the sum of their dimensions exceeds $\dim V$.

Problems ─────────────────────────────

1. Show explicitly that $\mathbb{R}^2$ is a vector space over $\mathbb{R}$ but not over $\mathbb{C}$.

2. Can a vector space have two different additive identities? Why?

3. Show that $\mathbb{F}_n[x]$, the collection of polynomials over a field $\mathbb{F}$ with degree at most n, is a vector space over $\mathbb{F}$ with respect to the ordinary addition and scalar multiplication of polynomials. Is $\mathbb{F}[x]$, the set of polynomials with any degree, a vector space over $\mathbb{F}$? What is the dimension of $\mathbb{F}_n[x]$ or $\mathbb{F}[x]$?

4. Determine whether or not the vectors $v_1 = 1 + x - 2x^2$, $v_2 = 2 + 5x - x^2$, and $v_3 = x + x^2$ in $\mathbb{F}_3[x]$ are linearly independent.

5. Show that $\{(1, i)^T, (i, -1)^T\}$ is a linearly independent subset of $\mathbb{C}^2$ over the real $\mathbb{R}$ but not over the complex $\mathbb{C}$.

6. Determine whether or not $\mathbb{R}^2$, with the operations

$$(x_1, y_1)^T + (x_2, y_2)^T = (x_1 x_2, y_1 y_2)^T$$

and

$$c(x_1, y_1)^T = (cx_1, cy_1)^T,$$

is a vector space over $\mathbb{R}$.

7. Let V be the set of all real numbers in the form

$$a + b\sqrt{2} + c\sqrt{5},$$

where a, b, and c are rational numbers. Show that V is a vector space over the rational number field $\mathbb{Q}$. Find $\dim V$ and a basis of V.

8. Let V be a vector space. If $u, v, w \in V$ are such that $au + bv + cw = 0$ for some scalars a, b, c, $ac \neq 0$, show that $\text{Span}\{u, v\} = \text{Span}\{v, w\}$.

9. Let V be a vector space over $\mathbb{F}$ and let W be a subset of V. Show that W is a subspace of V if and only if for all $u, v \in W$ and $c \in \mathbb{F}$

$$u + v \in W \quad \text{and} \quad cu \in W.$$

10. Is the set $\{(x, y)^T \in \mathbb{R}^2 : 2x - 3y = 0\}$ a subspace of $\mathbb{R}^2$? How about $\{(x, y)^T \in \mathbb{R}^2 : 2x - 3y = 1\}$? Give a geometric explanation.

11. Show that the set $\{(x, y - x, y)^T : x, y \in \mathbb{R}\}$ is a subspace of $\mathbb{R}^3$. Find the dimension and a basis of the subspace.

12. Show that if W is a subspace of vector space V of dimension n, then $\dim W \leq n$. Is it possible that $\dim W = n$ for a proper subspace W?

13. Let $\{u_1, \ldots, u_s\}$ and $\{v_1, \ldots, v_t\}$ be two sets of vectors. If $s > t$ and each u_i can be expressed as a linear combination of $v_1, \ldots, v_t$, show that $u_1, \ldots, u_s$ are linearly dependent.

14. Let V be a vector space over a field $\mathbb{F}$. Show that if $v \in V$ and $c \in \mathbb{F}$ such that $cv = 0$, then $c = 0$ or $v = 0$.

15. The sum $V_1 + V_2$ of two subspaces V_1 and V_2 of a finite-dimensional vector space V is called a *direct sum*, symbolized by $V_1 \oplus V_2$, if

$$v_1 + v_2 = 0, \; v_1 \in V_1, \; v_2 \in V_2 \quad \Rightarrow \quad v_1 = v_2 = 0.$$

Show that $V_1 + V_2$ is a direct sum if and only if

$$\dim(V_1 + V_2) = \dim V_1 + \dim V_2.$$

Conclude that if $\{u_1, \ldots, u_s\}$ is a basis for V_1 and $\{v_1, \ldots, v_t\}$ is a basis for V_2, then $\{u_1, \ldots, u_s, v_1, \ldots, v_t\}$ is a basis for $V_1 \oplus V_2$.

16. Let V_1 and V_2 be subspaces of a vector space of finite dimension such that $\dim(V_1 + V_2) = \dim(V_1 \cap V_2) + 1$. Show that $V_1 \subseteq V_2$ or $V_2 \subseteq V_1$.

17. Let S_1, S_2, and S_3 be subspaces of a vector space of dimension n. Show that

$$\dim(S_1 \cap S_2 \cap S_3) \geq \dim S_1 + \dim S_2 + \dim S_3 - 2n.$$

1.2 Matrices

An $m \times n$ *matrix* A over a field $\mathbb{F}$ ($\mathbb{C}$ or $\mathbb{R}$) is a rectangular array of m-rows and n-columns of entries in $\mathbb{F}$:

$$A = \begin{pmatrix} a_{11} & a_{12} & \cdots & a_{1n} \\ a_{21} & a_{22} & \cdots & a_{2n} \\ \vdots & \vdots & \vdots & \vdots \\ a_{m1} & a_{m2} & \cdots & a_{mn} \end{pmatrix}.$$

Such a matrix, written as $A = (a_{ij})$, is said to be of size $m \times n$.

The set of all $m \times n$ matrices over a field $\mathbb{F}$ is a vector space with respect to *matrix addition* by adding corresponding entries and to *scalar multiplication* by multiplying each entry of the matrix by the scalar. The dimension is mn, and the matrices with one entry equal to 1 and 0 entries elsewhere form a basis. In case of square matrices, that is, $m = n$, the dimension is n^2. We denote by $\mathbb{M}_n$ the set of all complex n-square matrices throughout the book.

The *product* AB of two matrices $A = (a_{ij})$ and $B = (b_{ij})$ is defined to be the matrix whose (i, j)-entry is given by

$$a_{i1}b_{1j} + a_{i2}b_{2j} + \cdots + a_{in}b_{nj}.$$

Thus, in order that AB make sense, the number of columns of A must equal the number of rows of B. Take, for example,

$$A = \begin{pmatrix} 1 & -1 \\ 0 & 2 \end{pmatrix}, \quad B = \begin{pmatrix} 3 & 4 & 5 \\ 6 & 0 & 8 \end{pmatrix}.$$

Then

$$AB = \begin{pmatrix} -3 & 4 & -3 \\ 12 & 0 & 16 \end{pmatrix}.$$

Note that BA is undefined.

Sometimes it is useful to write the matrix product AB, with $B = (b_1, b_2, \ldots, b_n)$, where the b_i are the column vectors of B, as

$$AB = (Ab_1, Ab_2, \ldots, Ab_n).$$

The *transpose* of an $m \times n$ matrix $A = (a_{ij})$ is an $n \times m$ matrix, denoted by A^T, whose (i,j)-entry is a_{ji}; and the *conjugate* of A is a matrix of the same size as A, symbolized by $\overline{A}$, whose (i,j)-entry is $\overline{a_{ij}}$. We denote the conjugate transpose $\overline{A}^T$ of A by A^*.

The $n \times n$ *identity* matrix I_n, or simply I, is the n-square matrix with all diagonal entries 1 and off-diagonal entries 0. A *scalar* matrix is a multiple of I, and a *zero* matrix 0 is a matrix with all entries 0.

A square complex matrix $A = (a_{ij})$ is said to be

diagonal	if $a_{ij} = 0$, $i \neq j$,
upper-triangular	if $a_{ij} = 0$, $i > j$,
symmetric	if $A^T = A$,
Hermitian	if $A^* = A$,
normal	if $A^*A = AA^*$,
unitary	if $A^*A = AA^* = I$, and
orthogonal	if $A^TA = AA^T = I$.

A *submatrix* of a given matrix is an array lying in specified subsets of the rows and columns of the given matrix. For example,

$$C = \begin{pmatrix} 1 & 2 \\ 3 & \frac{1}{4} \end{pmatrix}$$

is a submatrix of

$$A = \begin{pmatrix} 0 & 1 & 2 \\ i & 3 & \frac{1}{4} \\ \pi & \sqrt{3} & -1 \end{pmatrix}$$

lying in rows one and two and columns two and three.

If we write $B = (0, i)^T$, $D = (\pi)$, and $E = (\sqrt{3}, -1)$, then

$$A = \begin{pmatrix} B & C \\ D & E \end{pmatrix}.$$

The right-hand side is called a *partitioned* or *block* form of A.

The manipulation of partitioned matrices is a basic technique in matrix theory. One can perform addition and multiplication of (appropriately) partitioned matrices as with ordinary matrices.

For instance, if A, B, C, X, Y, U, V are n-square matrices, then

$$\begin{pmatrix} A & B \\ 0 & C \end{pmatrix} \begin{pmatrix} X & Y \\ U & V \end{pmatrix} = \begin{pmatrix} AX + BU & AY + BV \\ CU & CV \end{pmatrix}.$$

2×2 block matrices have appeared to be the most useful partitioned matrices. We shall emphasize the techniques for block matrices of this kind most of the time in this book.

Elementary row operations for matrices are those that

    i. interchange two rows,

    ii. multiply a row by a nonzero constant, or

    iii. add a multiple of a row to another row.

Elementary column operations are similarly defined, and similar operations on partitioned matrices will be discussed in Section 2.1.

An n-square matrix is called an *elementary matrix* if it can be obtained from I_n by a single elementary row operation. Elementary operations can be represented by elementary matrices. Let E be the elementary matrix by performing an elementary row (or column) operation on I_m (or I_n for column). If the same elementary row (or column) operation is performed on an $m \times n$ matrix A, then the resulting matrix from A via the elementary row (or column) operation is given by the product EA (or AE, respectively).

For instance, $A = \begin{pmatrix} 1 & 2 & 3 \\ 4 & 5 & 6 \end{pmatrix}$ is brought by elementary row and column operations into $\begin{pmatrix} 1 & 0 & 0 \\ 0 & 1 & 0 \end{pmatrix}$. Write in equations

$$R_3 R_2 R_1 \begin{pmatrix} 1 & 2 & 3 \\ 4 & 5 & 6 \end{pmatrix} C_1 C_2 = \begin{pmatrix} 1 & 0 & 0 \\ 0 & 1 & 0 \end{pmatrix},$$

where

$$R_1 = \begin{pmatrix} 1 & 0 \\ -4 & 1 \end{pmatrix}, \quad R_2 = \begin{pmatrix} 1 & 0 \\ 0 & -\frac{1}{3} \end{pmatrix}, \quad R_3 = \begin{pmatrix} 1 & -2 \\ 0 & 1 \end{pmatrix}$$

and

$$C_1 = \begin{pmatrix} 1 & 0 & 1 \\ 0 & 1 & 0 \\ 0 & 0 & 1 \end{pmatrix}, \quad C_2 = \begin{pmatrix} 1 & 0 & 0 \\ 0 & 1 & -2 \\ 0 & 0 & 1 \end{pmatrix}.$$

This generalizes as follows.

Theorem 1.2 *Let A be an $m \times n$ complex matrix. Then there exist $P \in \mathbb{M}_m$ and $Q \in \mathbb{M}_n$, products of elementary matrices, such that*

$$PAQ = \begin{pmatrix} I_r & 0 \\ 0 & 0 \end{pmatrix}. \tag{1.1}$$

The partitioned matrix in (1.1), written as $I_r \oplus 0$ and called a *direct sum* of I_r and 0, is uniquely determined by A. The size r of the identity I_r is the *rank* of A, denoted by $\operatorname{rank}(A)$. If $A = 0$, then $\operatorname{rank}(A) = 0$. Clearly $\operatorname{rank}(A^T) = \operatorname{rank}(\overline{A}) = \operatorname{rank}(A^*) = \operatorname{rank}(A)$.

An application of this theorem shows that the dimension of the *solution space* to the linear equation system $Ax = 0$, where A has n columns, is $n - \operatorname{rank}(A)$ (Problem 14). A notable fact about a linear equation system is that

$$Ax = 0 \quad \text{if and only if} \quad (A^*A)x = 0.$$

The *determinant* of a square matrix A, denoted by $\det A$, or $|A|$ if A is in a partitioned form, is a number associated to A. For

$$A = \begin{pmatrix} a & b \\ c & d \end{pmatrix}$$

the determinant is defined to be

$$\det A = \begin{vmatrix} a & b \\ c & d \end{vmatrix} = ad - bc.$$

In general, the determinant of $A \in \mathbb{M}_n$ may be defined inductively,

$$\det A = \sum_{j=1}^{n} (-1)^{1+j} a_{1j} \det A(1|j),$$

where $A(1|j)$ is a submatrix of A obtained by deleting row 1 and column j of A. This formula is referred to as the *Laplace expansion* along row 1. One can also expand a determinant along other rows or columns to get the same result.

The determinant of a matrix has the following properties:

a. the determinant changes sign if two rows are interchanged;

b. the determinant is unchanged if a constant multiple of one row is added to another row;

c. the determinant is a linear function of any row when all the other rows are held fixed.

Similar properties are true for columns. Two often-used facts are

$$\det(AB) = \det A \, \det B, \quad A, B \in \mathbb{M}_n,$$

and

$$\begin{vmatrix} A & B \\ 0 & C \end{vmatrix} = \det A \, \det C, \quad A \in \mathbb{M}_n, \, C \in \mathbb{M}_m.$$

A square matrix A is said to be *invertible* or *nonsingular* if there exists a matrix B of the same size such that

$$AB = BA = I.$$

Such a B, which can be proven to be unique, is called the *inverse* of A and denoted by A^{-1}. The inverse of A, when it exists, can be obtained from the *adjoint* of A, written as adj(A), whose (i,j)-entry is the *cofactor* of a_{ji}, that is, $(-1)^{j+i} \det A(j|i)$, where $A(j|i)$ stands for the submatrix of A obtained by deleting row j and column i. To be exact, in symbols,

$$A^{-1} = \frac{1}{\det A} \operatorname{adj}(A). \tag{1.2}$$

An effective way to find the inverse of a matrix A is to apply elementary row operations to the matrix (A, I) to get a matrix in the form (I, B). Then $B = A^{-1}$ (Problem 17).

If A is a square matrix, then $AB = I$ if and only if $BA = I$.

It is easy to see that rank $(A) = $ rank (PAQ) for invertible matrices P and Q of appropriate sizes (meaning that the involved operations for matrices can be performed). It can also be shown that the rank of A is the largest number of linearly independent columns (rows) of A. In addition, the rank of A is r if and only if there exists an r-square submatrix of A with nonzero determinant, but all $(r+1)$-square submatrices of A have determinant zero (Problem 18).

Theorem 1.3 *The following statements are equivalent for $A \in \mathbb{M}_n$:*

1. *A is invertible;*
2. *$AB = I$ or $BA = I$ for some $B \in \mathbb{M}_n$;*
3. *A is of rank n;*
4. *A is a product of elementary matrices;*
5. *$Ax = 0$ has only the trivial solution $x = 0$;*
6. *$Ax = b$ has a unique solution for each $b \in \mathbb{C}^n$;*
7. *$\det A \neq 0$;*
8. *the column vectors of A are linearly independent;*
9. *the row vectors of A are linearly independent.*

Problems

1. Find the rank of $\begin{pmatrix} 1 & 2 & 3 \\ 4 & 5 & 6 \\ 2 & 1 & 0 \end{pmatrix}$ by performing elementary operations.

2. Evaluate the determinants
$$\begin{vmatrix} 2 & -3 & 10 \\ 1 & 2 & -2 \\ 0 & 1 & -3 \end{vmatrix}, \quad \begin{vmatrix} 1+x & 1 & 1 \\ 1 & 1+y & 1 \\ 1 & 1 & 1+z \end{vmatrix}.$$

3. Show the 3×3 *Vandermonde* determinant identity
$$\begin{vmatrix} 1 & 1 & 1 \\ a_1 & a_2 & a_3 \\ a_1^2 & a_2^2 & a_3^2 \end{vmatrix} = (a_2 - a_1)(a_3 - a_1)(a_3 - a_2)$$

and evaluate the determinant
$$\begin{vmatrix} 1 & a & a^2 - bc \\ 1 & b & b^2 - ca \\ 1 & c & c^2 - ab \end{vmatrix}.$$

4. Does there exist a real matrix A of odd order such that $A^2 = -I$?

5. Let $A \in \mathbb{M}_n$. Show that $A^* + A$ is Hermitian and $A^* - A$ is normal.

6. Let A and B be complex matrices of appropriate sizes. Show that
    (a) $\overline{AB} = \overline{A}\,\overline{B}$,
    (b) $(AB)^T = B^T A^T$,
    (c) $(AB)^* = B^* A^*$,
    (d) $(AB)^{-1} = B^{-1} A^{-1}$ if A and B are invertible.

7. If a, b, c, and d are complex numbers such that $ad - bc \neq 0$, show that
$$\begin{pmatrix} a & b \\ c & d \end{pmatrix}^{-1} = \frac{1}{ad - bc} \begin{pmatrix} d & -b \\ -c & a \end{pmatrix}.$$

8. Find the inverse of each of the following matrices:
$$\begin{pmatrix} 1 & a & 0 \\ 0 & 1 & 0 \\ 0 & b & 1 \end{pmatrix}, \quad \begin{pmatrix} 1 & 1 & 0 \\ 0 & 1 & 1 \\ 0 & 0 & 1 \end{pmatrix}, \quad \begin{pmatrix} 1 & 1 & 0 \\ 1 & 1 & 1 \\ 0 & 1 & 1 \end{pmatrix}.$$

9. Compute for every positive integer k
$$\begin{pmatrix} 0 & 1 \\ 1 & 0 \end{pmatrix}^k, \quad \begin{pmatrix} 1 & 1 \\ 1 & 0 \end{pmatrix}^k, \quad \begin{pmatrix} 1 & 1 \\ 0 & 1 \end{pmatrix}^k.$$

10. Show that for any square matrices A and B of the same size
$$A^*A - B^*B = \frac{1}{2}((A+B)^*(A-B) + (A-B)^*(A+B)).$$

11. Show that any two of the following three properties imply the third:
    (a) $A = A^*$;   (b) $A^* = A^{-1}$;   (c) $A^2 = I$.

12. Let A be a square complex matrix. Show that
$$I - A^{m+1} = (I - A)(I + A + A^2 + \cdots + A^m).$$

13. Let A, B, C, and D be n-square complex matrices. Compute
$$\begin{pmatrix} A & A^* \\ A^* & A \end{pmatrix}^2 \quad \text{and} \quad \begin{pmatrix} A & B \\ C & D \end{pmatrix}\begin{pmatrix} D & -B \\ -C & A \end{pmatrix}.$$

14. Show that the solution set to the linear system $Ax = 0$ is a vector space of dimension $n - \text{rank}(A)$ for any $m \times n$ matrix A over $\mathbb{R}$ or $\mathbb{C}$.
15. Let $A, B \in \mathbb{M}_n$. If $AB = 0$, show that $\text{rank}(A) + \text{rank}(B) \leq n$.
16. Let A and B be complex matrices with the same number of columns. If $Bx = 0$ whenever $Ax = 0$, show that
$$\text{rank}(B) \leq \text{rank}(A), \quad \text{rank}\begin{pmatrix} A \\ B \end{pmatrix} = \text{rank}(A),$$
and that $B = CA$ for some matrix C. When is C invertible?
17. Let $A, B \in \mathbb{M}_n$. If $B(A, I) = (I, B)$, show that $B = A^{-1}$. Explain why A^{-1}, if it exists, can be obtained by row operations.
18. Show that the following statements are equivalent for $A \in \mathbb{M}_n$:
    (a) $PAQ = \begin{pmatrix} I_r & 0 \\ 0 & 0 \end{pmatrix}$ for some invertible matrices P and Q;
    (b) the largest number of column (row) vectors of A that constitute a linearly independent set is r;
    (c) A contains an $r \times r$ nonsingular submatrix, and every $(r+1)$-square submatrix has determinant zero.

    (Hint: View P and Q as sequences of elementary operations. Note that rank does not change under elementary operations.)
19. Prove Theorem 1.3.
20. Let S be the $n \times n$ *backward identity* matrix, that is,
$$S = \begin{pmatrix} 0 & 0 & \ldots & 0 & 1 \\ 0 & 0 & \ldots & 1 & 0 \\ \vdots & \vdots & \vdots & \vdots & \vdots \\ 0 & 1 & \ldots & 0 & 0 \\ 1 & 0 & \ldots & 0 & 0 \end{pmatrix}.$$
Show that $S^{-1} = S^T = S$; equivalently, $S^T = S$, $S^2 = I$, $S^T S = I$. What is $\det S$? When $n = 3$, compute SAS for $A \in \mathbb{M}_3$.
21. Let A and B be $n \times n$ matrices. Show that for any $n \times n$ matrix X
$$\text{rank}\begin{pmatrix} A & X \\ 0 & B \end{pmatrix} \geq \text{rank}(A) + \text{rank}(B).$$
Discuss the cases where $X = 0$ and $X = I$, respectively.

1.3 Linear Transformations and Eigenvalues

Let V and W be vector spaces over a field $\mathbb{F}$. A map $\mathcal{A}: V \mapsto W$ is called a *linear transformation* from V to W if for all $u, v \in V$, $c \in \mathbb{F}$

$$\mathcal{A}(u + v) = \mathcal{A}(u) + \mathcal{A}(v)$$

and

$$\mathcal{A}(cv) = c\mathcal{A}(v).$$

It is easy to check that $\mathcal{A}: \mathbb{R}^2 \mapsto \mathbb{R}^2$, defined by

$$\mathcal{A}(x_1, x_2)^T = (x_1 + x_2, x_1 - x_2)^T,$$

is a linear transformation and that the *differential operator* $\mathcal{D}_x$ from $C'[a, b]$, the set (space) of functions with continuous derivatives on the interval $[a, b]$, to $C[a, b]$, the set of continuous functions on $[a, b]$, defined by

$$\mathcal{D}_x(f) = \frac{df(x)}{dx}, \quad f \in C'[a, b],$$

is a linear transformation.

Let $\mathcal{A}$ be a linear transformation from V to W. The subset in W

$$\text{Im}(\mathcal{A}) = \{\mathcal{A}(v) : v \in V\}$$

is a subspace of W, called the *image* of $\mathcal{A}$, and the subset in V

$$\text{Ker}(\mathcal{A}) = \{v \in V : \mathcal{A}(v) = 0 \in W\}$$

is a subspace of V, called the *kernel* or *null space* of $\mathcal{A}$.

Theorem 1.4 *Let $\mathcal{A}$ be a linear transformation from a vector space V of dimension n to a vector space W. Then*

$$\dim \text{Im}(\mathcal{A}) + \dim \text{Ker}(\mathcal{A}) = n.$$

SEC. 1.3 LINEAR TRANSFORMATIONS AND EIGENVALUES 15

This is seen by taking a basis $\{u_1, \ldots, u_s\}$ for Ker$(\mathcal{A})$ and extending it to a basis $\{u_1, \ldots, u_s, v_1, \ldots, v_t\}$ for V, where $s + t = n$. It is easy to show that $\{\mathcal{A}(v_1), \ldots, \mathcal{A}(v_t)\}$ is a basis of Im$(\mathcal{A})$.

Given an $m \times n$ matrix A with entries in $\mathbb{F}$, one can always define a linear transformation $\mathcal{A}$ from $\mathbb{F}^n$ to $\mathbb{F}^m$ by

$$\mathcal{A}(x) = Ax, \quad x \in \mathbb{F}^n. \tag{1.3}$$

Conversely, linear transformations can be represented by matrices. Consider, for example, $\mathcal{A} : \mathbb{R}^2 \mapsto \mathbb{R}^3$ defined by

$$\mathcal{A}(x_1, x_2)^T = (3x_1, 2x_1 + x_2, -x_1 - 2x_2)^T.$$

Then $\mathcal{A}$ is a linear transformation. We may write in the form

$$\mathcal{A}(x) = Ax,$$

where

$$x = (x_1, x_2)^T, \quad A = \begin{pmatrix} 3 & 0 \\ 2 & 1 \\ -1 & -2 \end{pmatrix}.$$

Let $\mathcal{A}$ be a linear transformation from V to W. Once the bases for V and W have been chosen, $\mathcal{A}$ has a unique matrix representation A as in (1.3) determined by the images of the basis vectors of V under $\mathcal{A}$, and there is a one-to-one correspondence between the linear transformations and their matrices. A linear transformation may have different matrices under different bases. In what follows we show that these matrices are similar when $V = W$. Two square matrices A and B of the same size are said to be *similar* if $P^{-1}AP = B$ for some invertible matrix P.

Let $\mathcal{A}$ be a linear transformation on a vector space V with a basis $\{u_1, \ldots, u_n\}$. Since each $\mathcal{A}(u_i)$ is a vector in V, we may write

$$\mathcal{A}(u_i) = \sum_{j=1}^{n} a_{ji} u_j, \quad i = 1, \ldots, n, \tag{1.4}$$

and call $A = (a_{ij})$ the *matrix of $\mathcal{A}$ under the basis* $\{u_1, \ldots, u_n\}$.

Write (1.4) conventionally as

$$\mathcal{A}(u_1, \ldots, u_n) = (\mathcal{A}(u_1), \ldots, \mathcal{A}(u_n)) = (u_1, \ldots, u_n)A.$$

Let $v \in V$. If $v = x_1 u_1 + \cdots + x_n u_n$, then

$$\mathcal{A}(v) = \sum_{i=1}^{n} x_i \mathcal{A}(u_i) = (\mathcal{A}(u_1), \ldots, \mathcal{A}(u_n))x = (u_1, \ldots, u_n)Ax,$$

where x is the column vector $(x_1, \ldots, x_n)^T$. In case of $\mathbb{R}^n$ or $\mathbb{C}^n$ with the standard basis $u_1 = e_1, \ldots, u_n = e_n$, we have

$$\mathcal{A}(v) = Ax.$$

Let $\{v_1, \ldots, v_n\}$ also be a basis of V. Expressing each u_i as a linear combination of $v_1, \ldots, v_n$ gives an $n \times n$ matrix B such that

$$(u_1, \ldots, u_n) = (v_1, \ldots, v_n)B.$$

It can be shown (Problem 9) that B is invertible since $\{u_1, \ldots, u_n\}$ is a linearly independent set. It follows by using (1.4) that

$$\begin{aligned}\mathcal{A}(v_1, \ldots, v_n) &= \mathcal{A}((u_1, \ldots, u_n)B^{-1}) \\ &= (u_1, \ldots, u_n)AB^{-1} \\ &= (v_1, \ldots, v_n)(BAB^{-1}).\end{aligned}$$

This says the matrices of a linear transformation under different bases $\{u_1, \ldots, u_n\}$ and $\{v_1, \ldots, v_n\}$ are similar.

Given a linear transformation on a vector space, it is a central theme of linear algebra to find a basis of the vector space so that the matrix of a linear transformation is as simple as possible, in the sense that the matrix contains more zeros or has a particular structure. In the words of matrices, the given matrix is reduced to a canonical form via similarity. This will be discussed in Chapter 3.

Let $\mathcal{A}$ be a linear transformation on a vector space V over $\mathbb{C}$. A nonzero vector $v \in V$ is called an *eigenvector* of $\mathcal{A}$ belonging to an *eigenvalue* $\lambda \in \mathbb{C}$ if

$$\mathcal{A}(v) = \lambda v, \quad v \neq 0.$$

If, for example, $\mathcal{A}$ is defined on $\mathbb{R}^2$ by

$$\mathcal{A}(x, y)^T = (y, x)^T,$$

then $\mathcal{A}$ has two eigenvalues, 1 and -1. What are the eigenvectors?

SEC. 1.3   LINEAR TRANSFORMATIONS AND EIGENVALUES   17

If λ_1 and λ_2 are different eigenvalues of $\mathcal{A}$ with respective eigenvectors x_1 and x_2, then x_1 and x_2 are linearly independent, for if

$$l_1 x_1 + l_2 x_2 = 0 \qquad (1.5)$$

for some scalars l_1 and l_2, then applying $\mathcal{A}$ to both sides yields

$$l_1 \lambda_1 x_1 + l_2 \lambda_2 x_2 = 0. \qquad (1.6)$$

Multiplying both sides of (1.5) by λ_1, we have

$$l_1 \lambda_1 x_1 + l_2 \lambda_1 x_2 = 0. \qquad (1.7)$$

Subtracting (1.6) from (1.7) results in

$$l_2(\lambda_1 - \lambda_2)x_2 = 0.$$

It follows that $l_2 = 0$, and thus $l_1 = 0$ from (1.5).

This can be generalized by induction to the following statement:

If α_{i_j} are linearly independent eigenvectors corresponding to an eigenvalue λ_i, then the set of all eigenvectors α_{i_j} for these eigenvalues λ_i together is linearly independent. Simply put:

Theorem 1.5 *The eigenvectors belonging to different eigenvalues are linearly independent.*

If $\mathcal{A}$ happens to have n linearly independent eigenvectors belonging to (not necessarily distinct) eigenvalues $\lambda_1, \lambda_2, \ldots, \lambda_n$, then $\mathcal{A}$, under the basis formed by the corresponding eigenvectors, has a diagonal matrix representation

$$\begin{pmatrix} \lambda_1 & & & 0 \\ & \lambda_2 & & \\ & & \ddots & \\ 0 & & & \lambda_n \end{pmatrix}.$$

To find eigenvalues and eigenvectors, one needs to convert

$$\mathcal{A}(v) = \lambda v$$

under a basis into a linear equation system

$$Ax = \lambda x.$$

Therefore, the eigenvalues of $\mathcal{A}$ are those $\lambda \in \mathbb{F}$ such that

$$\det(\lambda I - A) = 0,$$

and the eigenvectors of $\mathcal{A}$ are the vectors whose coordinates under the basis are the solutions to the equation system $Ax = \lambda x$.

Suppose A is an $n \times n$ complex matrix. The polynomial in λ

$$\det(\lambda I_n - A) \tag{1.8}$$

is called the *characteristic polynomial* of A, and the zeros of the polynomial are called the *eigenvalues of A*. It follows that every n-square matrix has n eigenvalues over $\mathbb{C}$ (including repeated ones).

The *trace* of an n-square matrix A, denoted by $\operatorname{tr} A$, is defined to be the sum of the eigenvalues $\lambda_1, \ldots, \lambda_n$ of A, that is,

$$\operatorname{tr} A = \lambda_1 + \cdots + \lambda_n.$$

It is easy to see from (1.8) by expanding the determinant that

$$\operatorname{tr} A = a_{11} + \cdots + a_{nn}$$

and

$$\det A = \prod_{i=1}^{n} \lambda_i.$$

Problems ─────────────────────────────

1. Show that the map $\mathcal{A}$ from $\mathbb{R}^3$ to itself defined by

$$\mathcal{A}(x, y, z)^T = (x + y, x - y, z)^T$$

is a linear transformation. Find its matrix under the standard basis.

2. Find the dimensions of $\operatorname{Im}(\mathcal{A})$ and $\operatorname{Ker}(\mathcal{A})$, and find their bases for the linear transformation $\mathcal{A}$ on $\mathbb{R}^3$ defined by

$$\mathcal{A}(x, y, z)^T = (x - 2z, y + z, 0)^T.$$

3. Define a linear transformation $\mathcal{A}: \mathbb{R}^2 \mapsto \mathbb{R}^2$ by
$$\mathcal{A}(x,y)^T = (y,0)^T.$$
   (a) Find $\text{Im}(\mathcal{A})$ and $\text{Ker}(\mathcal{A})$.
   (b) Find a matrix representation of $\mathcal{A}$.
   (c) Verify that $\dim \mathbb{R}^2 = \dim \text{Im}(\mathcal{A}) + \dim \text{Ker}(\mathcal{A})$.
   (d) Is $\text{Im}(\mathcal{A}) + \text{Ker}(\mathcal{A})$ a direct sum?
   (e) Does $\mathbb{R}^2 = \text{Im}(\mathcal{A}) + \text{Ker}(\mathcal{A})$?

4. Find the eigenvalues and eigenvectors of the differential operator $\mathcal{D}_x$.

5. Find the eigenvalues and corresponding eigenvectors of the matrix
$$A = \begin{pmatrix} 1 & 4 \\ 2 & 3 \end{pmatrix}.$$

6. Let λ be an eigenvalue of $\mathcal{A}$ on a vector space V, and let
$$V_\lambda = \{v \in V : \mathcal{A}(v) = \lambda v\},$$
called the *eigenspace* of λ. Show that V_λ is an *invariant subspace* of V under $\mathcal{A}$, that is, it is a subspace and $\mathcal{A}(v) \in V_\lambda$ for every $v \in V_\lambda$.

7. Define linear transformations $\mathcal{A}$ and $\mathcal{B}$ on $\mathbb{R}^2$ by
$$\mathcal{A}(x,y)^T = (x+y,\ y)^T, \quad \mathcal{B}(x,y)^T = (x+y,\ x-y)^T.$$
Find all eigenvalues of $\mathcal{A}$ and $\mathcal{B}$ and their eigenspaces.

8. Let $p(x) = \det(xI - A)$ be the characteristic polynomial of matrix $A \in \mathbb{M}_n$. If λ is an eigenvalue of A such that $p(x) = (x-\lambda)^k q(x)$ for some polynomial $q(x)$ with $q(\lambda) \neq 0$, show that
$$k \geq \dim V_\lambda.$$

9. Let $\{u_1, \ldots, u_n\}$ and $\{v_1, \ldots, v_n\}$ be two bases of a vector space V. Show that there exists an invertible matrix B such that
$$(u_1, \ldots, u_n) = (v_1, \ldots, v_n)B.$$

10. Let $\{u_1, \ldots, u_n\}$ be a basis for a vector space V and let $\{v_1, \ldots, v_k\}$ be a set of vectors in V. If $v_i = \sum_{j=1}^n a_{ij} u_j$, $i = 1, \ldots, k$, show that
$$\dim \text{Span}\{v_1, \ldots, v_k\} = \text{rank}(A).$$

11. Show that similar matrices have the same trace and determinant.

12. Let v_1 and v_2 be eigenvectors of matrix A belonging to different eigenvalues λ_1 and λ_2, respectively. Show that $v_1 + v_2$ is not an eigenvector of A. How about $av_1 + bv_2$, $a, b \in \mathbb{R}$?

13. Let $A \in \mathbb{M}_n$ and let $S \in \mathbb{M}_n$ be nonsingular. If the first column of $S^{-1}AS$ is $(\lambda, 0, \ldots, 0)^T$, show that λ is an eigenvalue of A and that the first column of S is an eigenvector of A belonging to λ.

14. Let $x \in \mathbb{C}^n$. Find the eigenvalues and eigenvectors of the matrices
$$A_1 = xx^* \quad \text{and} \quad A_2 = \begin{pmatrix} 0 & x^* \\ x & 0 \end{pmatrix}.$$

15. If each row sum (i.e., the sum of entries in a row) of matrix A is 1, show that 1 is an eigenvalue of A.

16. If λ is an eigenvalue of $A \in \mathbb{M}_n$, show that λ^2 is an eigenvalue of A^2 and that if A is invertible, then λ^{-1} is an eigenvalue of A^{-1}.

17. A *minor* of a matrix $A \in \mathbb{M}_n$ is the determinant of a square submatrix of A. Show that
$$\det(\lambda I - A) = \lambda^n - \delta_1 \lambda^{n-1} + \delta_2 \lambda^{n-2} - \cdots + (-1)^n \det A,$$
where δ_i is the sum of all principal minors of order i, $i = 1, 2, \ldots, n-1$. (A principal minor is the determinant of a submatrix indexed by the same rows and columns, called a *principal submatrix*.)

18. A linear transformation $\mathcal{A}$ on a vector space V is said to be invertible if there exists a linear transformation $\mathcal{B}$ such that $\mathcal{AB} = \mathcal{BA} = \mathcal{I}$, the identity. If $\dim V < \infty$, show that the following are equivalent:

    (a) $\mathcal{A}$ is invertible;
    (b) if $\mathcal{A}(x) = 0$, then $x = 0$, that is, $\text{Ker}(\mathcal{A}) = \{0\}$;
    (c) if $\{u_1, \ldots, u_n\}$ is a basis, then so is $\{\mathcal{A}u_1, \ldots, \mathcal{A}u_n\}$;
    (d) $\mathcal{A}$ is one-to-one;
    (e) $\mathcal{A}$ is onto, that is, $\text{Im}(\mathcal{A}) = V$;
    (f) $\mathcal{A}$ has a nonsingular matrix representation under some basis.

19. Let $\mathcal{A}$ be a linear transformation on a vector space of dimension n with matrix representation A. Show that
$$\dim \text{Im}(\mathcal{A}) = \text{rank}(A) \quad \text{and} \quad \dim \text{Ker}(\mathcal{A}) = n - \text{rank}(A).$$

20. Let $\mathcal{A}$ and $\mathcal{B}$ be linear transformations on a finite-dimensional vector space V having the same image, that is, $\mathrm{Im}(\mathcal{A}) = \mathrm{Im}(\mathcal{B})$. If
$$V = \mathrm{Im}(\mathcal{A}) \oplus \mathrm{Ker}(\mathcal{A}) = \mathrm{Im}(\mathcal{B}) \oplus \mathrm{Ker}(\mathcal{B}),$$
does it follow that $\mathrm{Ker}(\mathcal{A}) = \mathrm{Ker}(\mathcal{B})$?

21. Consider the vector space $\mathbb{F}[x]$ of all polynomials over $\mathbb{F}(=\mathbb{R} \text{ or } \mathbb{Q})$. For $f(x) = a_n x^n + a_{n-1} x^{n-1} + \cdots + a_1 x + a_0 \in \mathbb{F}[x]$, define
$$\mathcal{S}(f(x)) = \frac{a_n}{n+1} x^{n+1} + \frac{a_{n-1}}{n} x^n + \cdots + \frac{a_1}{2} x^2 + a_0 x$$
and
$$\mathcal{T}(f(x)) = n a_n x^{n-1} + (n-1) a_{n-1} x^{n-2} + \cdots + a_1.$$
Compute $\mathcal{ST}$ and $\mathcal{TS}$. Does $\mathcal{ST} = \mathcal{TS}$?

22. Define $\mathcal{P} : \mathbb{C}^n \mapsto \mathbb{C}^n$ by $\mathcal{P}(x) = (0, 0, x_3, \ldots, x_n)^T$. Show that $\mathcal{P}$ is a linear transformation and $\mathcal{P}^2 = \mathcal{P}$. What is $\mathrm{Ker}(\mathcal{P})$?

23. Let $\mathcal{A}$ be a linear transformation on a finite-dimensional vector space V, and let W be a subspace of V. Denote
$$\mathcal{A}(W) = \{\mathcal{A}(w) : w \in W\}.$$
Show that $\mathcal{A}(W)$ is a subspace of V. Furthermore, show that
$$\dim(\mathcal{A}(W)) + \dim(\mathrm{Ker}(\mathcal{A}) \cap W) = \dim W.$$

24. Let V be a vector space of dimension n over $\mathbb{C}$ and let $\{u_1, \ldots, u_n\}$ be a basis of V. For $x = x_1 u_1 + \cdots + x_n u_n \in V$, define
$$\mathcal{T}(x) = (x_1, \ldots, x_n)^T \in \mathbb{C}^n,$$
Show that $\mathcal{T}$ is an *isomorphism*, or $\mathcal{T}$ is one-to-one, onto, and satisfies
$$\mathcal{T}(ax + by) = a\mathcal{T}(x) + b\mathcal{T}(y), \quad x, y \in V, \ a, b \in \mathbb{C}.$$

25. Let V be the vector space of all sequences
$$(c_1, c_2, \ldots), \quad c_i \in \mathbb{C}, \ i = 1, 2, \ldots.$$
Define a linear transformation on V by
$$\mathcal{S}(c_1, c_2, \cdots) = (0, c_1, c_2, \cdots).$$
Show that $\mathcal{S}$ has no eigenvalues. Moreover, if we define
$$\mathcal{S}^*(c_1, c_2, c_3, \cdots) = (c_2, c_3, \cdots),$$
then $\mathcal{S}^*\mathcal{S}$ is the identity, but $\mathcal{S}\mathcal{S}^*$ is not.

1.4 Inner Product Spaces

A vector space V over field $\mathbb{C}$ or $\mathbb{R}$ is called an *inner product space* if it is equipped with an *inner product* $(\,,\,)$ satisfying for all $u, v, w \in V$ and scalar c

1. $(u, u) \geq 0$, and $(u, u) = 0$ if and only if $u = 0$,
2. $(u + v, w) = (u, w) + (v, w)$,
3. $(cu, v) = c(u, v)$, and
4. $\overline{(u, v)} = (v, u)$.

$\mathbb{C}^n$ is an inner product space over $\mathbb{C}$ with the inner product

$$(x, y) = y^* x = \overline{y_1}\, x_1 + \cdots + \overline{y_n}\, x_n.$$

The *Cauchy-Schwarz inequality* for an inner product space is one of the most useful inequalities in mathematics.

Theorem 1.6 (Cauchy-Schwarz Inequality) *Let V be an inner product space. Then for all x and y in V,*

$$|(x, y)|^2 \leq (x, x)(y, y).$$

Equality holds if and only if x and y are linearly dependent.

The proof of this can be done in a number of different ways. The most common proof is to consider the quadratic function in t

$$(x + ty, x + ty)$$

and to derive the inequality from the nonnegative discriminant. One may also obtain the inequality from $(z, z) \geq 0$ by setting

$$z = y - \frac{(y, x)}{(x, x)} x, \quad x \neq 0,$$

and showing that $(z, x) = 0$ and then $(z, z) = (z, y) \geq 0$.

A matrix proof is left as an exercise for the reader (Problem 12).

For any vector x in an inner product space, the positive square root of (x, x) is called the *length* or *norm* of the vector x and is denoted by $\|x\|$, that is,

$$\|x\| = \sqrt{(x, x)}.$$

Thus, the Cauchy-Schwarz inequality is rewritten as

$$|(x, y)| \leq \|x\| \, \|y\|.$$

Theorem 1.7 *For all vectors x and y in an inner product space,*

i. $\|x\| \geq 0$; ii. $\|cx\| = |c| \|x\|, c \in \mathbb{C}$; iii. $\|x + y\| \leq \|x\| + \|y\|$.

The last inequality is referred to as the *triangle inequality*.

A *unit* vector is a vector whose length is 1. For any nonzero vector u, $\frac{1}{\|u\|} u$ is a unit vector. Two vectors x and y are said to be *orthogonal* if $(x, y) = 0$. An *orthogonal set* is a set in which any two of the vectors are orthogonal. Such a set is further said to be *orthonormal* if every vector in the set is of length 1.

For example, $\{v_1, v_2\}$ is an orthonormal set in $\mathbb{R}^2$, where

$$v_1 = \left(\frac{1}{\sqrt{2}}, \frac{1}{\sqrt{2}}\right), \quad v_2 = \left(\frac{1}{\sqrt{2}}, -\frac{1}{\sqrt{2}}\right).$$

The column (row) vectors of a unitary matrix are orthonormal.

Let S be a subset of an inner product space V. Denote by $S^\perp$ the collection of the vectors in V that are orthogonal to all vectors in S, that is,

$$S^\perp = \{v \in V : (v, s) = 0 \text{ for all } s \in S\}.$$

It is easy to see that $S^\perp$ is a subspace of V. If S contains only one element, say x, we simply use $x^\perp$ for $S^\perp$. For two subsets S_1 and S_2, if $(x, y) = 0$ for all $x \in S_1$ and $y \in S_2$, we write $S_1 \perp S_2$.

As we saw in the first section, a set of linearly independent vectors of a vector space of finite dimension can be extended to a basis for the vector space. Likewise a set of orthogonal vectors of an inner product space can be extended to an orthogonal basis of the space. The same is true for a set of orthonormal vectors. Consider $\mathbb{C}^n$, for instance. Let u_1 be a unit vector in $\mathbb{C}^n$. Pick a unit vector u_2 in $u_1^\perp$ if $n \geq 2$. Then u_1 and u_2 are orthonormal. Now if $n \geq 3$, let u_3 be a unit vector in $(\text{Span}\{u_1, u_2\})^\perp$ (equivalently, $(u_1, u_3) = 0$ and $(u_2, u_3) = 0$). Then u_1, u_2, u_3 are orthonormal. Continuing this way, one obtains an orthonormal basis for the inner product space. We summarize this as a theorem for $\mathbb{C}^n$, which will be freely and frequently used in the future.

Theorem 1.8 *If $u_1, \ldots, u_k$ are k linearly independent vectors in $\mathbb{C}^n$, $1 \leq k < n$, then there exist $n - k$ vectors $u_{k+1}, \ldots, u_n$ in $\mathbb{C}^n$ such that the matrix*

$$P = (u_1, \ldots, u_k, u_{k+1}, \ldots, u_n)$$

is invertible. Furthermore, if $u_1, \ldots, u_k$ are orthonormal, then there exist $n - k$ vectors $u_{k+1}, \ldots, u_n$ in $\mathbb{C}^n$ such that the matrix

$$U = (u_1, \ldots, u_k, u_{k+1}, \ldots, u_n)$$

is unitary. In particular, for any unit vector u in $\mathbb{C}^n$, there exists a unitary matrix that contains u as its first column.

If $\{u_1, \ldots, u_n\}$ is an orthonormal basis of an inner product space V over $\mathbb{C}$, and if x and y are two vectors in V expressed as

$$x = x_1 u_1 + \cdots + x_n u_n, \quad y = y_1 u_1 + \cdots + y_n u_n,$$

then $x_i = (x, u_i)$, $y_i = (y, u_i)$ for $i = 1, \ldots, n$,

$$\|x\| = |x_1|^2 + \cdots + |x_n|^2 \tag{1.9}$$

and

$$(x, y) = \overline{y_1}\, x_1 + \cdots + \overline{y_n}\, x_n. \tag{1.10}$$

For $A \in \mathbb{M}_n$ and with the standard basis $e_1, \ldots, e_n$ of $\mathbb{C}^n$, we have

$$\text{tr}\, A = \sum_{i=1}^{n} (Ae_i, e_i)$$

and for $x \in \mathbb{C}^n$

$$(Ax, x) = x^* A x = \sum_{i,j=1}^{n} a_{ij} \overline{x_i} x_j.$$

Upon computation, we have

$$(Ax, y) = (x, A^* y)$$

and, with $\operatorname{Im} A = \{Ax : x \in \mathbb{C}^n\}$ and $\operatorname{Ker} A = \{x \in \mathbb{C}^n : Ax = 0\}$,

$$\operatorname{Ker} A^* = (\operatorname{Im} A)^\perp, \quad \operatorname{Im} A^* = (\operatorname{Ker} A)^\perp. \qquad (1.11)$$

$\mathbb{M}_n$ is an inner product space with the inner product

$$(A, B)_{\mathbb{M}} = \operatorname{tr}(B^* A), \quad A, B \in \mathbb{M}_n.$$

It is immediate by the Cauchy-Schwarz inequality that

$$|\operatorname{tr}(AB)|^2 \le \operatorname{tr}(A^* A)\operatorname{tr}(B^* B)$$

and that

$$\operatorname{tr}(A^* A) = 0 \quad \text{if and only if} \quad A = 0.$$

We end this section by presenting an inequality of the angles between vectors. This inequality is intuitive and obvious in $\mathbb{R}^2$ and $\mathbb{R}^3$. The good part of this theorem is the idea in its proof of reducing the problem to $\mathbb{R}^2$ or $\mathbb{R}^3$.

Let V be an inner product space over $\mathbb{R}$. For any nonzero vectors x and y, define the *angle* between x and y by

$$\angle_{x,y} = \cos^{-1} \frac{(x,y)}{\|x\|\|y\|}.$$

Theorem 1.9 *For any nonzero vectors x, y, and z in an inner product space V over $\mathbb{R}$,*

$$\angle_{x,z} \le \angle_{x,y} + \angle_{y,z}.$$

Equality occurs if and only if $y = ax + bz$, $a, b \ge 0$.

PROOF. Since the inequality involves only the vectors x, y, and z, we may focus on the subspace $\text{Span}\{x, y, z\}$, which has dimension at most 3. We can further choose an orthonormal basis (a unit vector in case of dimension one) for this subspace. Let x, y, and z have coordinates α, β, and γ under the basis, respectively. Then the inequality holds if and only if it holds for real vectors α, β, and γ, due to (1.9) and (1.10). Thus, the problem is reduced to $\mathbb{R}$, $\mathbb{R}^2$, or $\mathbb{R}^3$ depending on whether the dimension of $\text{Span}\{x, y, z\}$ is 1, 2, or 3, respectively. For $\mathbb{R}$, the assertion is trivial. For $\mathbb{R}^2$ or $\mathbb{R}^3$, a simple graph will do the job. ∎

Problems

1. If V is an inner product space over $\mathbb{C}$, show that for $x, y \in V$, $c \in \mathbb{C}$
$$(x, cy) = \bar{c}(x, y) \quad \text{and} \quad (x, y)(y, x) = |(x, y)|^2.$$

2. Find all vectors in $\mathbb{R}^2$ (with the usual inner product) that are orthogonal to $(1, 1)^T$. Is $(1, 1)^T$ a unit vector?

3. Show that in an inner product space over $\mathbb{R}$ or $\mathbb{C}$
$$(x, y) = 0 \quad \Rightarrow \quad \|x + y\|^2 = \|x\|^2 + \|y\|^2$$
and that the converse is true over $\mathbb{R}$ but not over $\mathbb{C}$.

4. Show that for any two vectors x and y in an inner product space
$$\|x - y\|^2 + \|x + y\|^2 = 2(\|x\|^2 + \|y\|^2).$$

5. Is [,] defined as $[x, y] = x_1 y_1 + \cdots + x_n y_n$ an inner product for $\mathbb{C}^n$?

6. For what diagonal $D \in \mathbb{M}_n$ is $[x, y] = y^* D x$ an inner product for $\mathbb{C}^n$?

7. Let $\{u_1, \ldots, u_n\}$ be an orthonormal basis of an inner product space V. Show that for $x \in V$
$$(u_i, x) = 0, \quad i = 1, \ldots, n, \quad \Leftrightarrow \quad x = 0,$$
and that for $x, y \in V$
$$(u_i, x) = (u_i, y), \quad i = 1, \ldots, n, \quad \Leftrightarrow \quad x = y.$$

8. Let $\{v_1, \ldots, v_n\}$ be a set of vectors in an inner product space V. Denote the matrix with entries (v_i, v_j) by G. Show that $\det G = 0$ if and only if $v_1, \ldots, v_n$ are linearly dependent.

9. Let A be an n-square complex matrix. Show that
$$\operatorname{tr}(AX) = 0 \text{ for every } X \in \mathbb{M}_n \quad \Leftrightarrow \quad A = 0.$$

10. Let $A \in \mathbb{M}_n$. Show that for any unit vector $x \in \mathbb{C}^n$,
$$|x^*Ax|^2 \leq x^*A^*Ax.$$

11. Let $A = (a_{ij})$ be a complex matrix with rows a_i. Show that
$$\operatorname{tr}(A^*A) = \operatorname{tr}(AA^*) = \sum_{i,j} |a_{ij}|^2.$$

12. Use the fact that $\operatorname{tr}(A^*A) \geq 0$ with equality if and only if $A = 0$ to show the Cauchy-Schwarz inequality for $\mathbb{C}^n$ by taking $A = xy^* - yx^*$.

13. Let A and B be complex matrices of the same size. Show that
$$|\operatorname{tr}(AB)| \leq \left(\operatorname{tr}(A^*A)\operatorname{tr}(B^*B)\right)^{\frac{1}{2}} \leq \frac{1}{2}\left(\operatorname{tr}(A^*A) + \operatorname{tr}(B^*B)\right).$$

14. Let A and B be n-square complex matrices. For every $x \in \mathbb{C}^n$, if
$$(Ax, x) = (Bx, x),$$
does it follow that $A = B$? What if $x \in \mathbb{R}^n$?

15. Show that for any n-square complex matrix A
$$\mathbb{C}^n = \operatorname{Im} A \oplus (\operatorname{Im} A)^\perp = \operatorname{Im} A \oplus \operatorname{Ker} A^*$$
and that $\mathbb{C}^n = \operatorname{Im} A \oplus \operatorname{Ker} A$ if and only if $\operatorname{rank}(A^2) = \operatorname{rank}(A)$.

16. Let θ_i and λ_i be positive numbers and $\sum_{i=1}^n \theta_i = 1$. Show that
$$1 \leq \left(\sum_{i=1}^n \theta_i \lambda_i\right)\left(\sum_{i=1}^n \theta_i \lambda_i^{-1}\right).$$

17. Let $\{u_1, \ldots, u_n\}$ be an orthonormal basis of an inner product space V. Show that $x_1, \ldots, x_k$ in V are pairwise orthogonal if and only if
$$\sum_{k=1}^n (x_i, u_k)\overline{(x_j, u_k)} = 0, \quad i \neq j.$$

18. If $\{u_1, \ldots, u_k\}$ is an orthonormal set in an inner product space V of dimension n, show that $k \leq n$ and for any $x \in V$

$$\|x\|^2 \geq \sum_{i=1}^{k} |(x, u_i)|^2.$$

19. Let $\{u_1, \ldots, u_n\}$ and $\{v_1, \ldots, v_n\}$ be two orthonormal bases of an inner product space. Show that there exists a unitary U such that

$$(u_1, \ldots, u_n) = (v_1, \ldots, v_n)U.$$

20. Prove or disprove for unit vectors u, v, w in an inner product space

$$|(u, w)| \leq |(u, v)| + |(v, w)|.$$

21. Let V_1 and V_2 be subsets of an inner product space V. Show that

$$V_1 \subseteq V_2 \quad \Rightarrow \quad V_2^\perp \subseteq V_1^\perp.$$

22. Let V_1 and V_2 be subspaces of an inner product space V. Show that

$$(V_1 + V_2)^\perp = V_1^\perp \cap V_2^\perp$$

and

$$(V_1 \cap V_2)^\perp = V_1^\perp + V_2^\perp.$$

23. Let V_1 and V_2 be subspaces of an inner product space V of dimension n. If $\dim V_1 > \dim V_2$, show that there exists a subspace V_3 such that

$$V_3 \subset V_1, \quad V_3 \perp V_2, \quad \dim V_3 \geq \dim V_1 - \dim V_2.$$

Give a geometric explanation in $\mathbb{R}^3$.

24. Let A and B be $m \times n$ complex matrices. Show that

$$\operatorname{Im} A \perp \operatorname{Im} B \quad \Leftrightarrow \quad A^*B = 0.$$

25. Let A be an n-square complex matrix. Show that for any $x, y \in \mathbb{C}^n$

$$4(Ax, y) = (As, s) - (At, t) + i(Au, u) - i(Av, v),$$

where $s = x + y$, $t = x - y$, $u = x + iy$, $v = x - iy$.

26. Show that for any nonzero x and y in $\mathbb{C}^n$

$$\|x - y\| \geq \frac{1}{2}(\|x\| + \|y\|) \left\| \frac{1}{\|x\|}x - \frac{1}{\|y\|}y \right\|.$$

CHAPTER 2

Partitioned Matrices

INTRODUCTION This chapter is devoted to the techniques of partitioned (block) matrices. Topics include elementary operations, determinants, and inverses of partitioned matrices. We begin with the elementary operations of block matrices, followed by discussions of the inverse and rank of the sum and product of matrices. We then present four different proofs of the theorem that the products AB and BA of matrices A and B of sizes $m \times n$ and $n \times m$, respectively, have the same nonzero eigenvalues. At the end of this chapter we discuss the often-used matrix technique of continuity argument.

2.1 Elementary Operations of Partitioned Matrices

The manipulation of partitioned matrices is a basic tool in matrix theory. The techniques for manipulating partitioned matrices resemble those for ordinary numerical matrices. We begin by considering a 2×2 matrix
$$\begin{pmatrix} a & b \\ c & d \end{pmatrix}, \quad a,\ b,\ c,\ d \in \mathbb{C}.$$
An application of an elementary row operation, say, adding the second row multiplied by -3 to the first row, can be represented by the

matrix multiplication

$$\begin{pmatrix} 1 & -3 \\ 0 & 1 \end{pmatrix} \begin{pmatrix} a & b \\ c & d \end{pmatrix} = \begin{pmatrix} a - 3c & b - 3d \\ c & d \end{pmatrix}.$$

Note that the determinant of the matrix does not change.

Elementary row or column operations for matrices play an important role in elementary linear algebra. These operations (page 8) can be generalized to partitioned matrices as follows:

I. interchange two (block) rows (columns),

II. multiply a (block) row (column) from the left (right) by a nonsingular matrix of appropriate size, and

III. multiply a (block) row (column) by a matrix from the left (right), then add it to another row (column).

Write in matrices, say, for type III elementary row operations,

$$\begin{pmatrix} A & B \\ C & D \end{pmatrix} \to \begin{pmatrix} A & B \\ C + XA & D + XB \end{pmatrix},$$

where $A \in \mathbb{M}_m$, $D \in \mathbb{M}_n$, and X is $n \times m$. Note that A is multiplied by X from the left (when row operations are performed).

Generalized elementary matrices are those obtained by applying a single elementary operation to the identity matrix. For instance,

$$\begin{pmatrix} 0 & I_m \\ I_n & 0 \end{pmatrix} \quad \text{and} \quad \begin{pmatrix} I_m & 0 \\ X & I_n \end{pmatrix}$$

are generalized elementary matrices of type I and type III.

Theorem 2.1 *Let G be the generalized elementary matrix obtained by performing an elementary row (column) operation on I. If that same elementary row (column) operation is performed on a block matrix A, then the resulting matrix is given by the product GA (AG).*

PROOF. We show the case of 2×2 partitioned matrices since we shall deal with this type of partitioned matrix most of the time. An argument for the general case is similar.

SEC. 2.1   ELEMENTARY OPERATIONS OF PARTITIONED MATRICES   31

Let A, B, C, and D be matrices, where A and D are m- and n-square, respectively. Suppose we apply a type III operation, say, adding the first row times an $n \times m$ matrix E from the left to the second row, to the matrix

$$\begin{pmatrix} A & B \\ C & D \end{pmatrix}.$$

Then we have, by writing in equation,

$$\begin{pmatrix} A & B \\ C+EA & D+EB \end{pmatrix} = \begin{pmatrix} I_m & 0 \\ E & I_n \end{pmatrix} \begin{pmatrix} A & B \\ C & D \end{pmatrix}. \quad \blacksquare$$

As an application, if A is invertible, we can make the lower-left and upper-right submatrices 0 by subtracting the first row multiplied by CA^{-1} from the the second row, and by subtracting the first column multiplied by $A^{-1}B$ from the second column. In symbols,

$$\begin{pmatrix} A & B \\ C & D \end{pmatrix} \to \begin{pmatrix} A & B \\ 0 & D-CA^{-1}B \end{pmatrix} \to \begin{pmatrix} A & 0 \\ 0 & D-CA^{-1}B \end{pmatrix},$$

and in equation,

$$\begin{pmatrix} I_m & 0 \\ -CA^{-1} & I_n \end{pmatrix} \begin{pmatrix} A & B \\ C & D \end{pmatrix} \begin{pmatrix} I_m & -A^{-1}B \\ 0 & I_n \end{pmatrix}$$
$$= \begin{pmatrix} A & 0 \\ 0 & D-CA^{-1}B \end{pmatrix}. \quad (2.1)$$

Note that by taking determinants,

$$\begin{vmatrix} A & B \\ C & D \end{vmatrix} = \det A \, \det(D - CA^{-1}B).$$

The method of manipulating block matrices by elementary operations and the corresponding generalized elementary matrices as in (2.1) will be used repeatedly in this book.

For practice, we now consider expressing the block matrix

$$\begin{pmatrix} A & B \\ 0 & A^{-1} \end{pmatrix} \quad (2.2)$$

as a product of block matrices of the forms
$$\begin{pmatrix} I & X \\ 0 & I \end{pmatrix}, \quad \begin{pmatrix} I & 0 \\ Y & I \end{pmatrix}.$$

In other words, we want to get a matrix in the above form by performing type III operations on the block matrix in (2.2).

Add the first row of (2.2) times A^{-1} to the second row to get
$$\begin{pmatrix} A & B \\ I & A^{-1} + A^{-1}B \end{pmatrix}.$$

Add the second row multiplied by $I - A$ to the first row to get
$$\begin{pmatrix} I & A^{-1} + A^{-1}B - I \\ I & A^{-1} + A^{-1}B \end{pmatrix}.$$

Subtract the first row from the second row to get
$$\begin{pmatrix} I & A^{-1} + A^{-1}B - I \\ 0 & I \end{pmatrix},$$

which is in the desired form. Putting these steps in identity, we have
$$\begin{pmatrix} I & 0 \\ -I & I \end{pmatrix} \begin{pmatrix} I & I - A \\ 0 & I \end{pmatrix} \begin{pmatrix} I & 0 \\ A^{-1} & I \end{pmatrix} \begin{pmatrix} A & B \\ 0 & A^{-1} \end{pmatrix}$$
$$= \begin{pmatrix} I & A^{-1} + A^{-1}B - I \\ 0 & I \end{pmatrix}.$$

Therefore,
$$\begin{pmatrix} A & B \\ 0 & A^{-1} \end{pmatrix} = \begin{pmatrix} I & 0 \\ A^{-1} & I \end{pmatrix}^{-1} \begin{pmatrix} I & I - A \\ 0 & I \end{pmatrix}^{-1}$$
$$\times \begin{pmatrix} I & 0 \\ -I & I \end{pmatrix}^{-1} \begin{pmatrix} I & A^{-1} + A^{-1}B - I \\ 0 & I \end{pmatrix}$$
$$= \begin{pmatrix} I & 0 \\ -A^{-1} & I \end{pmatrix} \begin{pmatrix} I & A - I \\ 0 & I \end{pmatrix}$$
$$\times \begin{pmatrix} I & 0 \\ I & I \end{pmatrix} \begin{pmatrix} I & A^{-1} + A^{-1}B - I \\ 0 & I \end{pmatrix}$$

is a product of type III generalized elementary matrices.

Problems

1. Let $E = E[i(c) \to j]$ denote the elementary matrix obtained from I_n by adding row i times c to row j.
   (a) Show that $E^* = E[j(\bar{c}) \to i]$.
   (b) Show that $E^{-1} = E[i(-c) \to j]$.
   (c) How is E obtained via an elementary column operation?

2. Let X be any $n \times m$ complex matrix. Show that
$$\begin{pmatrix} I_m & 0 \\ X & I_n \end{pmatrix}^{-1} = \begin{pmatrix} I_m & 0 \\ -X & I_n \end{pmatrix}.$$

3. Show that for any n-square complex matrix X
$$\begin{pmatrix} X & I_n \\ I_n & 0 \end{pmatrix}^{-1} = \begin{pmatrix} 0 & I_n \\ I_n & -X \end{pmatrix}.$$
Does it follow that
$$\begin{pmatrix} 0 & I_m \\ I_n & 0 \end{pmatrix}^{-1} = \begin{pmatrix} 0 & I_m \\ I_n & 0 \end{pmatrix}?$$

4. Show that every 2×2 matrix of determinant 1 is the product of some matrices of the following types, with $y \neq 0$:
$$\begin{pmatrix} 1 & 0 \\ x & 1 \end{pmatrix}, \begin{pmatrix} 1 & x \\ 0 & 1 \end{pmatrix}, \begin{pmatrix} 0 & 1 \\ 1 & 0 \end{pmatrix}, \begin{pmatrix} y & 0 \\ 0 & 1 \end{pmatrix}, \begin{pmatrix} 1 & 0 \\ 0 & y \end{pmatrix}.$$

5. Let X and Y be matrices with the same number of rows. Multiply
$$\begin{pmatrix} X & Y \\ 0 & 0 \end{pmatrix} \begin{pmatrix} X^* & 0 \\ Y^* & 0 \end{pmatrix} \quad \text{and} \quad \begin{pmatrix} X^* & 0 \\ Y^* & 0 \end{pmatrix} \begin{pmatrix} X & Y \\ 0 & 0 \end{pmatrix}.$$

6. Let X and Y be complex matrices of the same size. Verify that
$$\begin{pmatrix} I + XX^* & X + Y \\ X^* + Y^* & I + Y^*Y \end{pmatrix} = \begin{pmatrix} I & X \\ Y^* & I \end{pmatrix} \begin{pmatrix} I & Y \\ X^* & I \end{pmatrix}$$
$$= \begin{pmatrix} X & I \\ I & Y^* \end{pmatrix} \begin{pmatrix} X^* & I \\ I & Y \end{pmatrix}$$
$$= \begin{pmatrix} X \\ I \end{pmatrix}(X^*, I) + \begin{pmatrix} I \\ Y^* \end{pmatrix}(I, Y).$$

7. Let $a_1, a_2, \ldots, a_n$ be complex numbers, $a = (-a_2, \ldots, -a_n)$, and

$$A = \begin{pmatrix} 0 & I_{n-1} \\ -a_1 & a \end{pmatrix}.$$

Find $\det A$. Show that A is invertible if $a_1 \neq 0$ and that

$$A^{-1} = \begin{pmatrix} \frac{1}{a_1}a & -\frac{1}{a_1} \\ I_{n-1} & 0 \end{pmatrix}.$$

8. Let $a_1, a_2, \ldots, a_n$ be nonzero complex numbers. Find

$$\begin{pmatrix} 0 & a_1 & 0 & \ldots & 0 \\ 0 & 0 & a_2 & \ldots & 0 \\ \vdots & \vdots & \vdots & \vdots & \vdots \\ 0 & 0 & 0 & \ldots & a_{n-1} \\ a_n & 0 & 0 & \ldots & 0 \end{pmatrix}^{-1}.$$

9. Show that the following two block matrices commute:

$$\begin{pmatrix} I_m & A \\ 0 & I_n \end{pmatrix}, \quad \begin{pmatrix} I_m & B \\ 0 & I_n \end{pmatrix}.$$

10. Show that a generalized elementary matrix $\begin{pmatrix} I & X \\ 0 & I \end{pmatrix}$ can be written as the product of the same type of elementary matrices with only one nonzero off-diagonal entry. (Hint: Consider how to get x_{ij} in the matrix by a type iii elementary operation on page 8.)

11. Let A and B be nonsingular matrices. Find $\begin{pmatrix} A & C \\ 0 & B \end{pmatrix}^{-1}$.

12. Let A and B be m- and n-square matrices, respectively. Show that

$$\begin{pmatrix} A & * \\ 0 & B \end{pmatrix}^k = \begin{pmatrix} A^k & * \\ 0 & B^k \end{pmatrix},$$

where the $*$ are some matrices, and that if A and B are invertible,

$$\begin{pmatrix} A & * \\ 0 & B \end{pmatrix}^{-1} = \begin{pmatrix} A^{-1} & * \\ 0 & B^{-1} \end{pmatrix}.$$

13. Let A and B be $n \times n$ complex matrices. Show that
$$\begin{vmatrix} 0 & A \\ B & 0 \end{vmatrix} = (-1)^n \det A \det B$$
and that if A and B are invertible, then
$$\begin{pmatrix} 0 & A \\ B & 0 \end{pmatrix}^{-1} = \begin{pmatrix} 0 & B^{-1} \\ A^{-1} & 0 \end{pmatrix}.$$

14. Let A and B be $n \times n$ matrices. Apply elementary operations to $\begin{pmatrix} A & 0 \\ I & B \end{pmatrix}$ to get $\begin{pmatrix} AB & 0 \\ B & I \end{pmatrix}$. Deduce $\det(AB) = \det A \det B$.

15. Let A and B be $n \times n$ matrices. Apply elementary operations to $\begin{pmatrix} I & A \\ B & I \end{pmatrix}$ to get $\begin{pmatrix} I - AB & 0 \\ 0 & I \end{pmatrix}$ and $\begin{pmatrix} I & 0 \\ 0 & I - BA \end{pmatrix}$. Conclude that $I - AB$ and $I - BA$ have the same rank for $A, B \in \mathbb{M}_n$.

16. Let A and B be $n \times n$ matrices. Apply elementary operations to $\begin{pmatrix} A & 0 \\ 0 & B \end{pmatrix}$ to get $\begin{pmatrix} A+B & B \\ B & B \end{pmatrix}$ and derive the rank inequality
$$\text{rank}\,(A+B) \leq \text{rank}\,(A) + \text{rank}\,(B).$$

17. Let A be a square complex matrix partitioned as
$$A = \begin{pmatrix} A_{11} & A_{12} \\ A_{21} & A_{22} \end{pmatrix}, \quad A_{11} \in \mathbb{M}_m,\ A_{22} \in \mathbb{M}_n.$$
Show that for any $B \in \mathbb{M}_m$
$$\begin{vmatrix} BA_{11} & BA_{12} \\ A_{21} & A_{22} \end{vmatrix} = \det B\ \det A$$
and for any $n \times m$ matrix C
$$\begin{vmatrix} A_{11} & A_{12} \\ A_{21} + CA_{11} & A_{22} + CA_{12} \end{vmatrix} = \det A.$$

18. Let A be a square complex matrix. Show that
$$\begin{vmatrix} I & A \\ A^* & I \end{vmatrix} = 1 - \sum M_1^* M_1 + \sum M_2^* M_2 - \sum M_3^* M_3 + \cdots,$$
where each M_k is a minor of order $k = 1, 2, \ldots$. (Hint: Reduce the left-hand side to $\det(I - A^*A)$ and use Problem 17 of Section 1.3.)

2.2 The Determinant and Inverse of Partitioned Matrices

Let M be a square complex matrix partitioned as

$$M = \begin{pmatrix} A & B \\ C & D \end{pmatrix},$$

where A and D are m- and n-square matrices, respectively. We discuss the determinants and inverses of matrices in this form. The results are fundamental and used almost everywhere in matrix theory, such as matrix computation and matrix inequalities. The methods of continuity and finding inverses deserve special attention.

Theorem 2.2 *Let M be a square matrix partitioned as above. Then*

$$\det M = \det A \, \det(D - CA^{-1}B), \quad \text{if } A \text{ is invertible,}$$

and

$$\det M = \det(AD - CB), \quad \text{if } AC = CA.$$

PROOF. When A^{-1} exists, it is easy to verify (see also (2.1)) that

$$\begin{pmatrix} I_m & 0 \\ -CA^{-1} & I_n \end{pmatrix} \begin{pmatrix} A & B \\ C & D \end{pmatrix} = \begin{pmatrix} A & B \\ 0 & D - CA^{-1}B \end{pmatrix}.$$

By taking determinants for both sides, we have

$$\det M = \begin{vmatrix} A & B \\ 0 & D - CA^{-1}B \end{vmatrix}$$
$$= \det A \, \det(D - CA^{-1}B).$$

For the second part, if A and C commute, then A, B, C, and D are of the same size. We show the identity by the so-called continuity argument method.

First consider the case where A is invertible. Following the above argument and using the fact that

$$\det(XY) = \det X \det Y$$

for any two square matrices X and Y of the same size, we have

$$\begin{aligned} \det M &= \det A \det(D - CA^{-1}B) \\ &= \det(AD - ACA^{-1}B) \\ &= \det(AD - CAA^{-1}B) \\ &= \det(AD - CB). \end{aligned}$$

Now assume that A is singular. Since $\det(A + \epsilon I)$ as a polynomial in ϵ has a finite number of zeros, we may choose $\delta > 0$ such that

$$\det(A + \epsilon I) \neq 0 \quad \text{whenever } 0 < \epsilon < \delta,$$

that is, $A + \epsilon I$ is invertible for $\epsilon \in (0, \delta)$. Denote

$$M_\epsilon = \begin{pmatrix} A + \epsilon I & B \\ C & D \end{pmatrix}.$$

Noticing further that $A + \epsilon I$ and C commute, we have

$$\det M_\epsilon = \det((A + \epsilon I)D - CB) \quad \text{whenever } 0 < \epsilon < \delta.$$

Observe that both sides of the above equation are continuous functions of ϵ. Letting $\epsilon \to 0^+$ gives that

$$\det M = \det(AD - CB). \blacksquare$$

Note that the identity need not be true if $AC \neq CA$.

We now turn our attention to the inverses of partitioned matrices.

Theorem 2.3 *Suppose that the partitioned matrix*

$$M = \begin{pmatrix} A & B \\ C & D \end{pmatrix}$$

is invertible and that the inverse is conformally partitioned as

$$M^{-1} = \begin{pmatrix} X & Y \\ U & V \end{pmatrix},$$

where A, D, X, and V are square matrices. Then

$$\det A = \det V \det M. \tag{2.3}$$

PROOF. The identity (2.3) follows immediately by taking the determinants of both sides of the matrix identity

$$\begin{pmatrix} A & B \\ C & D \end{pmatrix} \begin{pmatrix} I & Y \\ 0 & V \end{pmatrix} = \begin{pmatrix} A & 0 \\ C & I \end{pmatrix}. \blacksquare$$

Note that the identity matrices I in the proof may have different sizes and that A is singular if and only if V is singular.

Theorem 2.4 *Let M and M^{-1} be as defined in Theorem 2.3. If A is a nonsingular principal submatrix of M, then*

$$\begin{aligned} X &= A^{-1} + A^{-1}B(D - CA^{-1}B)^{-1}CA^{-1}, \\ Y &= -A^{-1}B(D - CA^{-1}B)^{-1}, \\ U &= -(D - CA^{-1}B)^{-1}CA^{-1}, \\ V &= (D - CA^{-1}B)^{-1}. \end{aligned}$$

PROOF. As we know from elementary linear algebra (Theorem 1.2), every invertible matrix can be written as a product of elementary matrices; so can M^{-1}. Furthermore, since

$$M^{-1}(M, I) = (I, M^{-1}),$$

this says that we can obtain the inverse of M by performing row operations on (M, I) to get (I, M^{-1}) (Problem 17, Section 1.2).

We now apply row operations to the augmented block matrix

$$\begin{pmatrix} A & B & I & 0 \\ C & D & 0 & I \end{pmatrix}.$$

Multiply row 1 by A^{-1} (from the left) to get

$$\begin{pmatrix} I & A^{-1}B & A^{-1} & 0 \\ C & D & 0 & I \end{pmatrix}.$$

Subtract row 1 multiplied by C from row 2 to get

$$\begin{pmatrix} I & A^{-1}B & A^{-1} & 0 \\ 0 & D - CA^{-1}B & -CA^{-1} & I \end{pmatrix}.$$

Sec. 2.2   The Determinant and Inverse of Partitioned Matrices

Multiply row 2 by $(D - CA^{-1}B)^{-1}$ (which exists–why?) to get

$$\begin{pmatrix} I & A^{-1}B & A^{-1} & 0 \\ 0 & I & -(D - CA^{-1}B)^{-1}CA^{-1} & (D - CA^{-1}B)^{-1} \end{pmatrix}.$$

By subtracting row 2 times $A^{-1}B$ from row 1, we get the inverse of the partitioned matrix M in the form

$$\begin{pmatrix} A^{-1} + A^{-1}B(D - CA^{-1}B)^{-1}CA^{-1} & -A^{-1}B(D - CA^{-1}B)^{-1} \\ -(D - CA^{-1}B)^{-1}CA^{-1} & (D - CA^{-1}B)^{-1} \end{pmatrix}.$$

By comparison to the inverse M^{-1}, we have X, Y, U, and V with the desired expressions in terms of A, B, C, and D. ■

A similar discussion for a nonsingular D implies

$$X = (A - BD^{-1}C)^{-1}.$$

It follows that

$$(A - BD^{-1}C)^{-1} = A^{-1} + A^{-1}B(D - CA^{-1}B)^{-1}CA^{-1}. \quad (2.4)$$

There is a direct proof of (2.4) in next section. Other similar identities can also be derived, and a different approach to the inverses of partitioned matrices will be shown in Section 4 of Chapter 6.

Problems ─────────────────────────────────

1. Let A be an n-square nonsingular matrix. Show that

$$(\det A)^{-1} \begin{vmatrix} a & \alpha \\ \beta & A \end{vmatrix} = a - \alpha A^{-1}\beta.$$

2. Refer to Theorem 2.2 and assume $AC = CA$. Does it follow that

$$\begin{vmatrix} A & B \\ C & D \end{vmatrix} = \det(AD - BC)?$$

3. For matrices A, B, C of appropriate sizes, evaluate the determinants

$$\begin{vmatrix} A & I_n \\ I_m & 0 \end{vmatrix}, \quad \begin{vmatrix} 0 & A \\ A^{-1} & 0 \end{vmatrix}, \quad \begin{vmatrix} 0 & A \\ B & C \end{vmatrix}.$$

4. Let A, B, and C be n-square complex matrices. Show that

$$\begin{vmatrix} I_n & A \\ B & C \end{vmatrix} = \det(C - BA).$$

5. Let A and B be $m \times n$ and $n \times m$ matrices, respectively. Show that

$$\begin{vmatrix} I_n & B \\ A & I_m \end{vmatrix} = \begin{vmatrix} I_m & A \\ B & I_n \end{vmatrix}$$

and conclude that

$$\det(I_m - AB) = \det(I_n - BA).$$

Is it true that

$$\operatorname{rank}(I_m - AB) = \operatorname{rank}(I_n - BA)?$$

6. Can any two of the following expressions be identical for general complex square matrices A, B, C, D?

$$\det(AD - CB), \quad \det(AD - BC), \quad \det(DA - CB), \quad \det(DA - BC),$$

$$\begin{vmatrix} A & B \\ C & D \end{vmatrix}.$$

7. If A is an invertible matrix, show that

$$\operatorname{rank}\begin{pmatrix} A & B \\ C & D \end{pmatrix} = \operatorname{rank}(A) + \operatorname{rank}(D - CA^{-1}B).$$

In particular,

$$\operatorname{rank}\begin{pmatrix} I_n & I_n \\ X & Y \end{pmatrix} = n + \operatorname{rank}(X - Y).$$

8. Does it follow from the identity (2.3) that any principal submatrix (A) of a singular matrix (M) is singular?

9. Find the determinant and the inverse of the $2m \times 2m$ block matrix

$$A = \begin{pmatrix} aI_m & bI_m \\ cI_m & dI_m \end{pmatrix}, \quad ad - bc \neq 0.$$

10. If U is a unitary matrix partitioned as $U = \begin{pmatrix} u & x \\ y & U_1 \end{pmatrix}$, where $u \in \mathbb{C}$, show that $|u| = |\det U_1|$. What if U is real orthogonal?

Sec. 2.2 The Determinant and Inverse of Partitioned Matrices

11. Find the inverses, if they exist, for the matrices

$$\begin{pmatrix} A & I \\ I & 0 \end{pmatrix}, \quad \begin{pmatrix} I & X \\ Y & Z \end{pmatrix}.$$

12. Let A be an n-square nonsingular matrix. Write

$$A = B + iC, \quad A^{-1} = F + iG,$$

where B, C, F, G are real matrices, and set

$$D = \begin{pmatrix} B & -C \\ C & B \end{pmatrix}.$$

Show that D is nonsingular and that the inverse of D is

$$\begin{pmatrix} F & -G \\ G & F \end{pmatrix}.$$

In addition, D is normal if A is normal, and orthogonal if A is unitary.

13. Let A and C be m- and n-square invertible matrices, respectively. Show that for any $m \times n$ matrix B and $n \times m$ matrix D,

$$\det(A + BCD) = \det A \det C \det(C^{-1} + DA^{-1}B).$$

What can be deduced from this identity if $A = I_m$ and $C = I_n$?

14. Let A and B be real square matrices of the same size. Show that

$$\begin{vmatrix} A & -B \\ B & A \end{vmatrix} = |\det(A + iB)|^2.$$

15. Let A and B be complex square matrices of the same size. Show that

$$\begin{vmatrix} A & B \\ B & A \end{vmatrix} = \det(A + B)\det(A - B).$$

Also show that the eigenvalues of the 2×2 block matrix on the left-hand side consist of those of $A + B$ and $A - B$.

16. Let A and B be n-square matrices. For any integer k (positive or negative if A is invertible), find the (1, 2)-block of the matrix

$$\begin{pmatrix} A & B \\ 0 & I \end{pmatrix}^k.$$

17. Let B and C be complex matrices with the same number of rows, and let $A = (B, C)$. Show that

$$\begin{vmatrix} I & B^* \\ B & AA^* \end{vmatrix} = \det(CC^*)$$

and that if $C^*B = 0$ then

$$\det(A^*A) = \det(B^*B)\det(C^*C).$$

18. Let $A \in \mathbb{M}_n$. Show that there exists a diagonal matrix D with diagonal entries ± 1 such that $\det(A + D) \neq 0$. (Hint: Show by induction.)

19. Consider the partitioned matrix $M = \begin{pmatrix} A & B \\ C & D \end{pmatrix}$. Assume that the inverses involved exist, and denote

$$S = D - CA^{-1}B, \quad T = A - BD^{-1}C.$$

Show that each of the following expressions is equal to M^{-1}:

(a) $\begin{pmatrix} A^{-1} + A^{-1}BS^{-1}CA^{-1} & -A^{-1}BS^{-1} \\ -S^{-1}CA^{-1} & S^{-1} \end{pmatrix}$;

(b) $\begin{pmatrix} T^{-1} & -T^{-1}BD^{-1} \\ -D^{-1}CT^{-1} & D^{-1} + D^{-1}CT^{-1}BD^{-1} \end{pmatrix}$;

(c) $\begin{pmatrix} T^{-1} & -A^{-1}BS^{-1} \\ -D^{-1}CT^{-1} & S^{-1} \end{pmatrix}$;

(d) $\begin{pmatrix} T^{-1} & (C - DB^{-1}A)^{-1} \\ (B - AC^{-1}D)^{-1} & S^{-1} \end{pmatrix}$;

(e) $\begin{pmatrix} I & -A^{-1}B \\ 0 & I \end{pmatrix} \begin{pmatrix} A^{-1} & 0 \\ 0 & S^{-1} \end{pmatrix} \begin{pmatrix} I & 0 \\ -CA^{-1} & I \end{pmatrix}$;

(f) $\begin{pmatrix} A^{-1} & 0 \\ 0 & 0 \end{pmatrix} + \begin{pmatrix} A^{-1}B \\ -I \end{pmatrix} S^{-1}(CA^{-1}, -I).$

20. Deduce the following inverse identities from the previous problem:

$$\begin{aligned}
(A - BD^{-1}C)^{-1} &= A^{-1} + A^{-1}B(D - CA^{-1}B)^{-1}CA^{-1} \\
&= -C^{-1}D(B - AC^{-1}D)^{-1} \\
&= -(C - DB^{-1}A)^{-1}DB^{-1} \\
&= C^{-1}D(D - CA^{-1}B)^{-1}CA^{-1} \\
&= A^{-1}B(D - CA^{-1}B)^{-1}DB^{-1}.
\end{aligned}$$

2.3 The Inverse of a Sum

The inverse of a product is the reverse product of the inverses when they exist. In symbols,
$$(AB)^{-1} = B^{-1}A^{-1}$$
provided that A and B are invertible matrices of the same size.

The analog for the sum or difference of matrices does not hold in general since, for any matrix A, $A - A = 0$ is not invertible.

Theorem 2.5 *Let $A \in \mathbb{M}_m$ and $B \in \mathbb{M}_n$ be nonsingular matrices and let C and D be $m \times n$ and $n \times m$ matrices, respectively. If the matrix $A + CBD$ is nonsingular, then*
$$(A + CBD)^{-1} = A^{-1} - A^{-1}C(B^{-1} + DA^{-1}C)^{-1}DA^{-1}. \qquad (2.5)$$

PROOF. This is immediate from identity (2.4) in the previous section. We present a direct proof by computation below. First note that $B^{-1} + DA^{-1}C$ is nonsingular, since, by Problem 5 of Section 2.2,
$$\begin{aligned}
\det(B^{-1} + DA^{-1}C) &= \det B^{-1} \det(I_n + BDA^{-1}C) \\
&= \det B^{-1} \det(I_m + A^{-1}CBD) \\
&= \det B^{-1} \det A^{-1} \det(A + CBD) \neq 0.
\end{aligned}$$

We now prove (2.5) by a direct verification:
$$\begin{aligned}
&(A + CBD)\Big(A^{-1} - A^{-1}C(B^{-1} + DA^{-1}C)^{-1}DA^{-1}\Big) \\
&= I_m - C(B^{-1} + DA^{-1}C)^{-1}DA^{-1} + CBDA^{-1} \\
&\quad - CBDA^{-1}C(B^{-1} + DA^{-1}C)^{-1}DA^{-1} \\
&= I_m - C\Big((B^{-1} + DA^{-1}C)^{-1} - B \\
&\quad + BDA^{-1}C(B^{-1} + DA^{-1}C)^{-1}\Big)DA^{-1} \\
&= I_m - C\Big((I_n + BDA^{-1}C)(B^{-1} + DA^{-1}C)^{-1} - B\Big)DA^{-1} \\
&= I_m - C\Big(B(B^{-1} + DA^{-1}C)(B^{-1} + DA^{-1}C)^{-1} - B\Big)DA^{-1} \\
&= I_m - C(B - B)DA^{-1} \\
&= I_m. \qquad \blacksquare
\end{aligned}$$

A great number of matrix identities involving inverses can be derived from (2.5). The following two are immediate when the involved inverses exist:

$$(A + B)^{-1} = A^{-1} - A^{-1}(B^{-1} + A^{-1})^{-1}A^{-1}$$

and

$$(A + UV^*)^{-1} = A^{-1} - A^{-1}U(I + V^*A^{-1}U)^{-1}V^*A^{-1}.$$

Problems

1. If $I + A$ is nonsingular, show that $(I + A)^{-1}$ and $I - A$ commute.

2. Let A and B be invertible matrices of the same size. Show that
$$A^{-1} = B^{-1} - A^{-1}(A - B)B^{-1}$$
and that
$$A^{-1} + B^{-1} = A^{-1}(A + B)B^{-1}.$$

3. Let $A \in \mathbb{M}_n$ be nonsingular. Show that for any $B \in \mathbb{M}_n$
$$(A + B)A^{-1}(A - B) = (A - B)A^{-1}(A + B).$$

4. Let A, B, and $A + B$ be invertible matrices. Show that
$$(A^{-1} + B^{-1})^{-1} = A - A(A + B)^{-1}A.$$

5. Let $A + B$ be a nonsingular matrix. Show that
$$A - A(A + B)^{-1}A = B - B(A + B)^{-1}B.$$

6. Let $A + B$ be a nonsingular matrix. Show that
$$(A + B)^{-1}A = I - (A + B)^{-1}B$$
and
$$A(A + B)^{-1} = I - B(A + B)^{-1}.$$

7. Let A be a square matrix. Show that if the involved inverses exist,
$$(\alpha I - A)^{-1} - (\beta I - A)^{-1} = (\beta - \alpha)(\alpha I - A)^{-1}(\beta I - A)^{-1}.$$

8. Derive (2.5) from (2.4).

9. If matrix A has no eigenvalue -1, show that
$$(I+A)^{-1} + (I+A^{-1})^{-1} = I.$$

10. Let A and B be $m \times n$ and $n \times m$ matrices, respectively. If $I_n + BA$ is nonsingular, show that $I_m + AB$ is nonsingular and that
$$(I_n + BA)^{-1}B = B(I_m + AB)^{-1}.$$

11. Show that for any complex matrix A
$$(I + A^*A)^{-1}A^*A = A^*A(I + A^*A)^{-1} = I - (I + A^*A)^{-1}.$$

12. Show that when the involved inverses exist,
$$(I - AB)^{-1} = I + A(I - BA)^{-1}B.$$
Conclude that if $I - AB$ is invertible, then so is $I - BA$. In particular,
$$(I + AA^*)^{-1} = I - A(I + A^*A)^{-1}A^*.$$

13. Let A and B be complex matrices. Show that
$$AB = A + B \quad \Rightarrow \quad AB = BA.$$

14. Show that for $x, y \in \mathbb{C}^n$
$$\mathrm{adj}(I - xy^*) = xy^* + (1 - y^*x)I.$$

15. Let $u, v \in \mathbb{C}^n$ with $v^*u \neq 0$. Write $v^*u = p^{-1} + q^{-1}$. Show that
$$(I - puv^*)^{-1} = I - quv^*.$$

16. Let u and v be column vectors in $\mathbb{C}^n$ such that $v^*A^{-1}u$ is not equal to -1. Show that $A + uv^*$ is invertible and, with $\delta = 1 + v^*A^{-1}u$,
$$(A + uv^*)^{-1} = A^{-1} - \delta^{-1}A^{-1}uv^*A^{-1}.$$

17. Let A be an n-square matrix and let P be a nonsingular matrix of the same size such that $P^{-1}AP$ is upper-triangular. Write
$$P^{-1}AP = D - U,$$
where D is a diagonal matrix and U is an upper-triangular matrix with 0 on the diagonal. Show that if A is invertible, then
$$A^{-1} = PD^{-1}\Big(I + UD^{-1} + (UD^{-1})^2 + \cdots + (UD^{-1})^{n-1}\Big)P^{-1}.$$

2.4 The Rank of Product and Sum

This section is concerned with the ranks of the product AB and the sum $A + B$ in terms of the ranks of matrices A and B.

Recall that the kernel, or the null space, and the image of an $m \times n$ matrix A, viewed as a linear transformation, are respectively

$$\text{Ker}\, A = \{x \in \mathbb{C}^n : Ax = 0\}, \quad \text{Im}\, A = \{Ax : x \in \mathbb{C}^n\}.$$

Theorem 2.6 (Sylvester) *Let A and B be complex matrices of sizes $m \times n$ and $n \times p$, respectively. Then*

$$\text{rank}\,(AB) = \text{rank}\,(B) - \dim(\text{Im}\, B \cap \text{Ker}\, A). \qquad (2.6)$$

In particular,

$$\text{rank}\,(A) + \text{rank}\,(B) - n \leq \text{rank}\,(AB) \leq \min\{\text{rank}\,(A), \text{rank}\,(B)\}.$$

PROOF. Recall from Theorem 1.4 that if $\mathcal{A}$ is a linear transformation on an n-dimensional vector space, then

$$\dim\, \text{Im}(\mathcal{A}) + \dim\, \text{Ker}(\mathcal{A}) = n.$$

Viewing A as a linear transformation on $\mathbb{C}^n$, we have (Problem 1)

$$\text{rank}\,(A) = \dim\, \text{Im}\, A.$$

For the rank of AB, we think of A as a linear transformation on the vector space $\text{Im}\, B$. Its image is then $\text{Im}(AB)$ and its null space is $\text{Im}\, B \cap \text{Ker}\, A$. We thus have

$$\dim\, \text{Im}(AB) + \dim(\text{Im}\, B \cap \text{Ker}\, A) = \dim\, \text{Im}\, B.$$

The identity (2.6) then follows.

For the inequalities, the second one is immediate from (2.6), and the first one is due to the fact that

$$\dim\, \text{Im}\, A + \dim(\text{Im}\, B \cap \text{Ker}\, A) \leq n. \quad \blacksquare$$

For the product of three matrices, we have

$$\operatorname{rank}(ABC) \geq \operatorname{rank}(AB) + \operatorname{rank}(BC) - \operatorname{rank}(B). \quad (2.7)$$

A pure matrix proof of (2.7) goes as follows: Note that

$$\operatorname{rank}\begin{pmatrix} 0 & X \\ Y & Z \end{pmatrix} \geq \operatorname{rank}(X) + \operatorname{rank}(Y)$$

for any matrix Z of appropriate size, and that equality holds if $Z = 0$. The inequality (2.7) then follows from the matrix identity

$$\begin{pmatrix} I & -A \\ 0 & I \end{pmatrix} \begin{pmatrix} 0 & AB \\ BC & B \end{pmatrix} \begin{pmatrix} I & 0 \\ -C & I \end{pmatrix} = \begin{pmatrix} -ABC & 0 \\ 0 & B \end{pmatrix}.$$

Theorem 2.7 *Let A and B be $m \times n$ matrices, and denote by C and D, respectively, the partitioned matrices*

$$C = (I_m, \ I_m), \quad D = \begin{pmatrix} A \\ B \end{pmatrix}.$$

Then

$$\begin{aligned} \operatorname{rank}(A + B) &= \operatorname{rank}(A) + \operatorname{rank}(B) - \dim(\operatorname{Im} D \cap \operatorname{Ker} C) \\ &\quad - \dim(\operatorname{Im} A^* \cap \operatorname{Im} B^*). \end{aligned} \quad (2.8)$$

In particular,

$$\operatorname{rank}(A + B) \leq \operatorname{rank}(A) + \operatorname{rank}(B).$$

PROOF. Write

$$A + B = (I_m, \ I_m)\begin{pmatrix} A \\ B \end{pmatrix} = CD.$$

Utilizing the previous theorem, we have

$$\operatorname{rank}(A + B) = \operatorname{rank}(D) - \dim(\operatorname{Im} D \cap \operatorname{Ker} C). \quad (2.9)$$

However,

$$\begin{aligned}
\operatorname{rank}(D) &= \operatorname{rank}(D^*) \\
&= \operatorname{rank}(A^*,\ B^*) \\
&= \dim\ \operatorname{Im}(A^*,\ B^*) \\
&= \dim(\operatorname{Im} A^* + \operatorname{Im} B^*) \\
&= \dim\ \operatorname{Im} A^* + \dim\ \operatorname{Im} B^* \\
&\quad - \dim(\operatorname{Im} A^* \cap \operatorname{Im} B^*) \\
&= \operatorname{rank}(A^*) + \operatorname{rank}(B^*) - \dim(\operatorname{Im} A^* \cap \operatorname{Im} B^*) \\
&= \operatorname{rank}(A) + \operatorname{rank}(B) - \dim(\operatorname{Im} A^* \cap \operatorname{Im} B^*).
\end{aligned}$$

Substituting this into (2.9) gives (2.8). ∎

Problems

1. Show that $\operatorname{rank}(A) = \dim \operatorname{Im} A$ for any complex matrix A.

2. If $B \in \mathbb{M}_n$ is invertible, show that $\operatorname{rank}(AB) = \operatorname{rank}(A)$ for every $m \times n$ matrix A. Is the converse true?

3. Is it true that the sum of two singular matrices is singular? How about the product?

4. Let A be an $m \times n$ matrix. Show that if $\operatorname{rank}(A) = m$, then there is an $n \times m$ matrix B such that $AB = I_m$, and that if $\operatorname{rank}(A) = n$, then there is an $m \times n$ matrix B such that $BA = I_n$.

5. Let A be an $m \times n$ matrix. Show that for any $s \times m$ matrix X with columns linearly independent and any $n \times t$ matrix Y with rows linearly independent,
$$\operatorname{rank}(A) = \operatorname{rank}(XA) = \operatorname{rank}(AY) = \operatorname{rank}(XAY).$$

6. For matrices A and B, show that if $\operatorname{rank}(AB) = \operatorname{rank}(B)$, then
$$ABX = ABY \quad \Leftrightarrow \quad BX = BY,$$
and if $\operatorname{rank}(AB) = \operatorname{rank}(A)$, then
$$XAB = YAB \quad \Leftrightarrow \quad XA = YA.$$

7. For any matrices A and B of the same size, show that
$$\operatorname{Im}(A, B) = \operatorname{Im} A + \operatorname{Im} B.$$

8. Let A be an $m \times n$ matrix. Show that for any $n \times m$ matrix B,
$$\dim \operatorname{Im} A + \dim \operatorname{Ker} A = \dim \operatorname{Im}(BA) + \dim \operatorname{Ker}(BA).$$

9. Let A and B be $m \times n$ and $n \times m$ matrices, respectively. Show that
$$\det(AB) = 0 \quad \text{if } m > n.$$

10. For matrices A and B of the same size, show that
$$|\operatorname{rank}(A) - \operatorname{rank}(B)| \leq \operatorname{rank}(A \pm B).$$

11. Let A and B be n-square complex matrices. Show that
$$\operatorname{rank}(AB - I) \leq \operatorname{rank}(A - I) + \operatorname{rank}(B - I).$$

12. Let A and B be $m \times n$ matrices such that $B^*A = 0$. Show that
$$\operatorname{rank}(A + B) = \operatorname{rank}(A) + \operatorname{rank}(B) \leq n.$$

13. Show that if $A \in \mathbb{M}_n$ and $A^2 = A$, then $\operatorname{rank}(A) + \operatorname{rank}(I_n - A) = n$.

14. Let A be an $m \times n$ matrix with rank n. If $m > n$, show that there is a matrix B of size $(m - n) \times m$ and a matrix C of size $m \times (m - n)$, both of rank $m - n$, such that $BA = 0$ and (A, C) is nonsingular.

15. Let $\mathcal{A}$ be a linear transformation on a finite-dimensional vector space. Show that
$$\operatorname{Ker}(\mathcal{A}) \subseteq \operatorname{Ker}(\mathcal{A}^2) \subseteq \operatorname{Ker}(\mathcal{A}^3) \subseteq \cdots$$
and that
$$\operatorname{Im}(\mathcal{A}) \supseteq \operatorname{Im}(\mathcal{A}^2) \supseteq \operatorname{Im}(\mathcal{A}^3) \supseteq \cdots.$$
Further show that there are finite proper inclusions in each chain.

16. If A is an $m \times n$ complex matrix, $\operatorname{Im}(A)$ is in fact the space spanned by the column vectors of A, called the *column space* of A and denoted by $\mathcal{C}(A)$. Similarly, the row vectors of A span the *row space*, symbolized by $\mathcal{R}(A)$. Let A and B be two matrices. Show that
$$\mathcal{C}(A) \subseteq \mathcal{C}(B) \iff A = BC,$$
for some matrix C, and
$$\mathcal{R}(A) \subseteq \mathcal{R}(B) \iff A = RB,$$
for some matrix R. (Note: $\mathcal{C}(A) = \operatorname{Im} A$ and $\mathcal{R}(A) = \operatorname{Im} A^T$.)

17. Let A and B be $m \times n$ matrices. Show that
$$\mathcal{C}(A+B) \subseteq \mathcal{C}(A) + \mathcal{C}(B)$$
and that the following statements are equivalent:
   (a) $\mathcal{C}(A) \subseteq \mathcal{C}(A+B)$;
   (b) $\mathcal{C}(B) \subseteq \mathcal{C}(A+B)$;
   (c) $\mathcal{C}(A+B) = \mathcal{C}(A) + \mathcal{C}(B)$.

18. Prove or disprove, for any n-square matrices A and B, that
$$\operatorname{rank}\begin{pmatrix} A \\ B \end{pmatrix} = \operatorname{rank}(A, B).$$

19. Let A and B be matrices of the same size. Show the rank inequalities
$$\operatorname{rank}(A+B) \leq \operatorname{rank}\begin{pmatrix} A \\ B \end{pmatrix} \leq \operatorname{rank}(A) + \operatorname{rank}(B)$$
and
$$\operatorname{rank}(A+B) \leq \operatorname{rank}(A,B) \leq \operatorname{rank}(A) + \operatorname{rank}(B)$$
by writing
$$A + B = (A, B)\begin{pmatrix} I \\ I \end{pmatrix} = (I, I)\begin{pmatrix} A \\ B \end{pmatrix}.$$
Additionally, show that
$$\operatorname{rank}\begin{pmatrix} A & B \\ C & D \end{pmatrix} \leq \operatorname{rank}(A) + \operatorname{rank}(B) + \operatorname{rank}(C) + \operatorname{rank}(D).$$

20. Let A, B, and C be complex matrices of the same size. Show that
$$\operatorname{rank}(A,B,C) \leq \operatorname{rank}(A,B) + \operatorname{rank}(B,C) - \operatorname{rank}(B).$$

2.5 Eigenvalues of AB and BA

For square matrices A and B of the same size, the product matrices AB and BA need not be equal, or even similar. For instance, if

$$A = \begin{pmatrix} 1 & 0 \\ 1 & 0 \end{pmatrix}, \quad B = \begin{pmatrix} 0 & 0 \\ 1 & 1 \end{pmatrix},$$

then

$$AB = \begin{pmatrix} 0 & 0 \\ 0 & 0 \end{pmatrix}, \quad \text{but } BA = \begin{pmatrix} 0 & 0 \\ 2 & 0 \end{pmatrix}.$$

Note that in this example both $AB = 0$ and $BA \neq 0$ have only repeated eigenvalue 0 (twice, referred to as *multiplicity* of 0). Is this a coincidence, or can we construct an example such that AB has only zero eigenvalues but BA has some nonzero eigenvalues?

The following theorem gives a negative answer to the question. This is a very important result in matrix theory.

Theorem 2.8 *Let A and B be $m \times n$ and $n \times m$ complex matrices, respectively. Then AB and BA have the same nonzero eigenvalues, counting multiplicity.*

PROOF 1. Use determinants. Notice that

$$\begin{pmatrix} I_m & -A \\ 0 & \lambda I_n \end{pmatrix} \begin{pmatrix} \lambda I_m & A \\ B & I_n \end{pmatrix} = \begin{pmatrix} \lambda I_m - AB & 0 \\ \lambda B & \lambda I_n \end{pmatrix}$$

and that

$$\begin{pmatrix} I_m & 0 \\ -B & \lambda I_n \end{pmatrix} \begin{pmatrix} \lambda I_m & A \\ B & I_n \end{pmatrix} = \begin{pmatrix} \lambda I_m & A \\ 0 & \lambda I_n - BA \end{pmatrix}.$$

By taking determinants and equating the right-hand sides, we obtain

$$\lambda^n \det(\lambda I_m - AB) = \lambda^m \det(\lambda I_n - BA). \tag{2.10}$$

Thus, $\det(\lambda I_m - AB) = 0$ if and only if $\det(\lambda I_n - BA) = 0$ when $\lambda \neq 0$. It is immediate that AB and BA have the same nonzero eigenvalues, including multiplicity (by factorization).

PROOF 2. Use matrix similarity. Consider the block matrix

$$\begin{pmatrix} 0 & 0 \\ B & 0 \end{pmatrix}.$$

Add the second row multiplied by A from the left to the first row

$$\begin{pmatrix} AB & 0 \\ B & 0 \end{pmatrix}.$$

Do the similar operation for columns to get

$$\begin{pmatrix} 0 & 0 \\ B & BA \end{pmatrix}.$$

Write, in equation,

$$\begin{pmatrix} I_m & A \\ 0 & I_n \end{pmatrix} \begin{pmatrix} 0 & 0 \\ B & 0 \end{pmatrix} = \begin{pmatrix} AB & 0 \\ B & 0 \end{pmatrix}$$

and

$$\begin{pmatrix} 0 & 0 \\ B & 0 \end{pmatrix} \begin{pmatrix} I_m & A \\ 0 & I_n \end{pmatrix} = \begin{pmatrix} 0 & 0 \\ B & BA \end{pmatrix}.$$

It follows that

$$\begin{pmatrix} I_m & A \\ 0 & I_n \end{pmatrix}^{-1} \begin{pmatrix} AB & 0 \\ B & 0 \end{pmatrix} \begin{pmatrix} I_m & A \\ 0 & I_n \end{pmatrix} = \begin{pmatrix} 0 & 0 \\ B & BA \end{pmatrix},$$

that is, matrices

$$\begin{pmatrix} AB & 0 \\ B & 0 \end{pmatrix} \quad \text{and} \quad \begin{pmatrix} 0 & 0 \\ B & BA \end{pmatrix}$$

are similar. Thus, matrices AB and BA have the same nonzero eigenvalues, counting multiplicity. (Are AB and BA similar?)

PROOF 3. Use continuity argument. We first deal with the case where $m = n$. If A is nonsingular, then

$$BA = A^{-1}(AB)A.$$

Thus, AB and BA are similar and have the same eigenvalues.

If A is singular, let δ be such a positive number that $\epsilon I + A$ is nonsingular for every ϵ, $0 < \epsilon < \delta$. Then

$$(\epsilon I + A)B \quad \text{and} \quad B(\epsilon I + A)$$

are similar and have the same characteristic polynomials. Therefore,

$$\det(\lambda I_n - (\epsilon I_n + A)B) = \det(\lambda I_n - B(\epsilon I_n + A)), \quad 0 < \epsilon < \delta.$$

Since both sides are continuous functions of ϵ, letting $\epsilon \to 0$ gives

$$\det(\lambda I_n - AB) = \det(\lambda I_n - BA).$$

Thus, AB and BA have the same eigenvalues.

For the case where $m \neq n$, assume $m < n$. Augment A and B by zero rows and zero columns, respectively, so that

$$A_1 = \begin{pmatrix} A \\ 0 \end{pmatrix}, \quad B_1 = (B, 0)$$

are n-square matrices. Then

$$A_1 B_1 = \begin{pmatrix} AB & 0 \\ 0 & 0 \end{pmatrix} \quad \text{and} \quad B_1 A_1 = BA.$$

It follows that $A_1 B_1$ and $B_1 A_1$, consequently AB and BA, have the same nonzero eigenvalues, counting multiplicity.

PROOF 4. Treat matrices as operators. It must be shown that if $\lambda I_m - AB$ is singular, so is $\lambda I_n - BA$, and vice versa. It may be assumed that $\lambda = 1$, by multiplying $\frac{1}{\lambda}$ otherwise.

If $I_m - AB$ is invertible, let $X = (I_m - AB)^{-1}$. We compute

$$\begin{aligned}
(I_n - BA)(I_n + BXA) &= I_n + BXA - BA - BABXA \\
&= I_n + (BXA - BABXA) - BA \\
&= I_n + B(I_m - AB)XA - BA \\
&= I_n + BA - BA \\
&= I_n.
\end{aligned}$$

Thus, $I_n - BA$ is invertible. This approach gives no information on the multiplicity of the nonzero eigenvalues. ∎

As a side product, using (2.10), we have (see also Problem 5 of Section 2.2), for any $m \times n$ matrix A and $n \times m$ matrix B,
$$\det(I_m + AB) = \det(I_n + BA).$$
Note that $I_m + AB$ is invertible if and only if $I_n + BA$ is invertible.

Problems

1. Show that $\operatorname{tr}(AB) = \operatorname{tr}(BA)$ for any $m \times n$ matrix A and $n \times m$ matrix B. In particular, $\operatorname{tr}(A^*A) = \operatorname{tr}(AA^*)$.

2. For any square matrices A and B of the same size, show that
$$\operatorname{tr}(A+B)^2 = \operatorname{tr} A^2 + 2\operatorname{tr}(AB) + \operatorname{tr} B^2.$$
Does it follow that $(A+B)^2 = A^2 + 2AB + B^2$?

3. Let A and B be square matrices of the same size. Show that
$$\det(AB) = \det(BA).$$
Does this hold if A and B are not square? Is it true that
$$\operatorname{rank}(AB) = \operatorname{rank}(BA)?$$

4. Let A and B be $m \times n$ and $n \times m$ complex matrices, respectively, with $m < n$. If the eigenvalues of AB are $\lambda_1, \ldots, \lambda_m$, what are the eigenvalues of BA?

5. Let A and B be $n \times n$ matrices. Show that for every integer $k \geq 1$
$$\operatorname{tr}(AB)^k = \operatorname{tr}(BA)^k.$$
Does
$$\operatorname{tr}(AB)^k = \operatorname{tr}(A^k B^k)?$$

6. Show that for any $x, y \in \mathbb{C}^n$
$$\det(I_n + xy^*) = 1 + y^*x.$$

7. Compute the determinant
$$\begin{vmatrix} 1+x_1y_1 & x_1y_2 & \cdots & x_1y_n \\ x_2y_1 & 1+x_2y_2 & \cdots & x_2y_n \\ \vdots & \vdots & \vdots & \vdots \\ x_ny_1 & x_ny_2 & \cdots & 1+x_ny_n \end{vmatrix}.$$

Eigenvalues of AB and BA

8. If A, B, and C are three complex matrices of appropriate sizes, show that ABC, CAB, and BCA have the same set of nonzero eigenvalues. Is it true that ABC and CBA have the same nonzero eigenvalues?

9. Do A^* and A have the same nonzero eigenvalues? How about A^*A and AA^*? Show by example that $\det(AA^*) \neq \det(A^*A)$ in general.

10. For any square matrices A and B of the same size, show that
$$A^2 + B^2 \quad \text{and} \quad \begin{pmatrix} A^2 & AB \\ BA & B^2 \end{pmatrix}$$
have the same nonzero eigenvalues. Further show that the latter one must have zero eigenvalues. How many of them?

11. Let $A \in \mathbb{M}_n$. Find the eigenvalues of the block matrix
$$\begin{pmatrix} A & A \\ A & A \end{pmatrix}$$
in terms of the eigenvalues of A.

12. Let A be an $m \times n$ complex matrix and let M be the block matrix
$$M = \begin{pmatrix} 0 & A \\ A^* & 0 \end{pmatrix}.$$

Show that

(a) M is a Hermitian matrix,

(b) the eigenvalues of M are
$$-\sigma_1, \ldots, -\sigma_r, \overbrace{0, \ldots, 0}^{m+n-2r}, \sigma_r, \ldots, \sigma_1,$$
where $\sigma_1 \geq \cdots \geq \sigma_r$ are the positive square roots of the nonzero eigenvalues of A^*A, called *singular values* of A,

(c) $\det M = \det(-A^*A) = (-1)^n |\det A|^2$ if A is n-square,

(d) $2\begin{pmatrix} 0 & A \\ A^* & 0 \end{pmatrix} = \begin{pmatrix} B & A \\ A^* & C \end{pmatrix} + \begin{pmatrix} -B & A \\ A^* & -C \end{pmatrix}$ for any B, C.

13. Let A and B be matrices of sizes $m \times n$ and $n \times m$, respectively. Do AB and BA have the same nonzero singular values?

2.6 The Continuity Argument

One of the most frequently used techniques in matrix theory is the *continuity argument*. A good example of this is to show, as we saw in the previous section, that matrices AB and BA have the same set of eigenvalues when A and B are both square matrices of the same size, by first considering the case where A is invertible and concluding that AB and BA are similar:

$$AB = A(BA)A^{-1}.$$

If A is singular, consider $A + \epsilon I$. Choose $\delta > 0$ such that $A + \epsilon I$ is invertible for all ϵ, $0 < \epsilon < \delta$. Thus, $(A + \epsilon I)B$ and $B(A + \epsilon I)$ have the same set of eigenvalues for every $\epsilon \in (0, \delta)$.

Equate the characteristic polynomials to get

$$\det(\lambda I - (A + \epsilon I)B) = \det(\lambda I - B(A + \epsilon I)), \ \ 0 < \epsilon < \delta.$$

Since both sides are continuous functions of ϵ, letting $\epsilon \to 0^+$ gives

$$\det(\lambda I - AB) = \det(\lambda I - BA).$$

Thus, AB and BA have the same eigenvalues.

The proof was done in three steps:
1. Show that the assertion is true for the nonsingular A.
2. Replace singular A by nonsingular $A + \epsilon I$.
3. Use continuity of a function in ϵ to get the desired conclusion.

A similar argument applies to the proof of the identity for adjoint matrices (Problem 3):

$$\mathrm{adj}(AB) = \mathrm{adj}(B)\,\mathrm{adj}(A).$$

This is certainly an effective approach for many matrix problems. The setting in which the technique is used is rather important. Sometimes the result for nonsingular matrices may be invalid for the singular case. Here is an example of this kind, for which the continuity argument fails.

Theorem 2.9 *Let C and D be n-square matrices such that*
$$CD^T + DC^T = 0.$$
If D is nonsingular, then for any n-square matrices A and B
$$\begin{vmatrix} A & B \\ C & D \end{vmatrix} = \det(AD^T + BC^T).$$
The identity is invalid in general if D is singular.

PROOF. It is easy to verify that
$$\begin{pmatrix} A & B \\ C & D \end{pmatrix} \begin{pmatrix} D^T & 0 \\ C^T & I \end{pmatrix} = \begin{pmatrix} AD^T + BC^T & B \\ 0 & D \end{pmatrix}.$$
Taking determinants of both sides results in the desired identity.

For an example of the singular case, we take A, B, C, and D to be, respectively,
$$\begin{pmatrix} 1 & 0 \\ 0 & 0 \end{pmatrix}, \begin{pmatrix} 0 & 0 \\ 0 & 1 \end{pmatrix}, \begin{pmatrix} 0 & 1 \\ 0 & 0 \end{pmatrix}, \begin{pmatrix} 0 & 0 \\ 1 & 0 \end{pmatrix},$$
where D is singular. It is easy to see by a simple computation that the determinantal identity does not hold. ∎

We shall see in Chapter 6 that some matrix inequalities hold for positive definite matrices but not for positive semidefinite matrices.

Problems

1. Why did the continuity argument fail Theorem 2.9?
2. Let C and D be real matrices such that $CD^T + DC^T = 0$. Show that if C is *skew-symmetric* (i.e., $C^T = -C$), then so is DC.
3. Use a continuity argument to show that for any $A, B \in \mathbb{M}_n$
$$\mathrm{adj}(AB) = \mathrm{adj}(B)\,\mathrm{adj}(A).$$
4. Show that $A_\epsilon = P_\epsilon J_\epsilon P_\epsilon^{-1}$ if $\epsilon \neq 0$, where
$$A_\epsilon = \begin{pmatrix} \epsilon & 0 \\ 1 & 0 \end{pmatrix}, \quad P_\epsilon = \begin{pmatrix} 0 & \epsilon \\ 1 & 1 \end{pmatrix}, \quad J_\epsilon = \begin{pmatrix} 0 & 0 \\ 0 & \epsilon \end{pmatrix}.$$
What happens to the identity if $\epsilon \to 0$? Is A_0 similar to J_0?

5. Explain why rank $(A^2) \leq$ rank (A). Discuss whether a continuity argument can be applied to show the inequality.

6. Show that the eigenvalues of A are independent of ϵ, where
$$A = \begin{pmatrix} \epsilon - 1 & -1 \\ \epsilon^2 - \epsilon + 1 & -\epsilon \end{pmatrix}.$$

7. Denote by $\sigma_{\max}$ and $\sigma_{\min}$, respectively, the largest and the smallest singular values of the matrix A, where
$$A = \begin{pmatrix} 1 & \epsilon \\ \epsilon & 1 \end{pmatrix}, \quad \epsilon > 0.$$

Show that
$$\lim_{\epsilon \to 0^+} \frac{\sigma_{\max}}{\sigma_{\min}} = +\infty.$$

8. Let A be a nonsingular matrix with $A^{-1} = B = (b_{ij})$. Show that the b_{ij} are continuous functions of a_{ij}, the entries of A, and that
$$\lim_{\epsilon \to 0} \left(A(\epsilon)\right)^{-1} = A^{-1}$$
if $\lim_{\epsilon \to 0} A(\epsilon) = A$ entrywise. Conclude that
$$\lim_{\epsilon \to 0} (A - \epsilon I)^{-1} = A^{-1}$$
and for any $m \times n$ matrix X and $n \times m$ matrix Y, independent of ϵ,
$$\lim_{\epsilon \to 0} \begin{pmatrix} I_m & \epsilon X \\ \epsilon Y & I_n \end{pmatrix}^{-1} = I_{m+n}.$$

9. Let n be a positive number and x be a real number. Let
$$A = \begin{pmatrix} 1 & -\frac{x}{n} \\ \frac{x}{n} & 1 \end{pmatrix}.$$

Show that
$$\lim_{x \to 0} \left(\lim_{n \to \infty} \frac{1}{x}(I - A^n) \right) = \begin{pmatrix} 0 & 1 \\ -1 & 0 \end{pmatrix}.$$

(Hint: $A = cP$ for some constant c and orthogonal matrix P.)

· ———— ⊙ ———— ·

CHAPTER 3

Matrix Polynomials and Canonical Forms

INTRODUCTION This chapter is devoted to matrix decompositions. The main studies are on the Schur decomposition, spectral decomposition, singular value decomposition, Jordan decomposition, and numerical range. Attention is also paid to the polynomials that annihilate matrices, especially the minimal and characteristic polynomials, and to the similarity of a complex matrix to a real matrix.

3.1 Commuting Matrices

Matrices do not commute in general. Any square matrix A, however, commutes with polynomials in A.

A question arises: If a matrix B commutes with A, is it true that B can be expressed as a polynomial in A? The answer is negative, by taking A to be the identity matrix I and B to be a nondiagonal matrix of the same size as I. For some sorts of matrices, nevertheless, we have the following result.

Theorem 3.1 *Let all the eigenvalues of $A \in \mathbb{M}_n$ be distinct. If $B \in \mathbb{M}_n$ commutes with A, then B can be expressed uniquely as a polynomial in A with degree no more than $n - 1$.*

PROOF. To begin, recall from Theorem 1.5 that the eigenvectors belonging to different eigenvalues are linearly independent. Thus, a matrix with distinct eigenvalues has a set of linearly independent eigenvectors that form a basis of $\mathbb{C}^n$.

Let $u_1, u_2, \ldots, u_n$ be the eigenvectors corresponding to the eigenvalues $\lambda_1, \lambda_2, \ldots, \lambda_n$ of A, respectively. Set

$$P = (u_1, u_2, \ldots, u_n).$$

Then P is an invertible matrix and

$$P^{-1}AP = \text{diag}(\lambda_1, \lambda_2, \ldots, \lambda_n).$$

Let

$$P^{-1}AP = C \quad \text{and} \quad P^{-1}BP = D.$$

It follows from $AB = BA$ that $CD = DC$.

Since the diagonal entries of C are distinct, D must be a diagonal matrix too (Problem 1). Let $D = \text{diag}(\mu_1, \mu_2, \ldots, \mu_n)$.

Consider the linear equation system of unknowns $x_0, x_1, \ldots, x_{n-1}$:

$$x_0 + \lambda_1 x_1 + \cdots + \lambda_1^{n-1} x_{n-1} = \mu_1,$$

$$x_0 + \lambda_2 x_1 + \cdots + \lambda_2^{n-1} x_{n-1} = \mu_2,$$

$$\vdots$$

$$x_0 + \lambda_n x_1 + \cdots + \lambda_n^{n-1} x_{n-1} = \mu_n.$$

Since the coefficient matrix is a Vandermonde matrix that is nonsingular when $\lambda_1, \lambda_2, \ldots, \lambda_n$ are distinct (see Problem 3 of Section 1.2 or Theorem 4.9 in the next chapter), the equation system has a unique solution, say, $(a_0, a_1, \ldots, a_{n-1})$.

Define a polynomial with degree no more than $n-1$ by

$$p(x) = a_0 + a_1 x + a_2 x^2 + \cdots + a_{n-1} x^{n-1}.$$

It follows that

$$p(\lambda_i) = \mu_i, \quad i = 1, 2, \ldots, n.$$

It is immediate that $p(A) = B$ since $p(C) = D$ and that this polynomial $p(x)$ is unique for the solution to the system is unique. ∎

Such a method of finding a polynomial $p(x)$ of the given pairs is referred to as *interpolation*.

Theorem 3.2 *Let A and B be square matrices of the same size. If $AB = BA$, then there exists a unitary matrix U such that U^*AU and U^*BU are both upper-triangular.*

PROOF. We use induction on n. If $n = 1$, we have nothing to show. Suppose that the assertion is true for $n - 1$.

For the case of n, we consider matrices as linear transformations. Note that if A is a linear transformation on a finite-dimensional vector space V over $\mathbb{C}$, then A has at least one eigenvector in V, for

$$Ax = \lambda x, \quad x \neq 0, \quad \text{if and only if} \quad \det(\lambda I - A) = 0,$$

which has a solution in $\mathbb{C}$.

For each eigenvalue μ of B, consider the eigenspace of B

$$V_\mu = \{v \in \mathbb{C}^n : Bv = \mu v\}.$$

If A and B commute, then for every $v \in V_\mu$,

$$B(Av) = (BA)v = (AB)v = A(Bv) = A(\mu v) = \mu(Av).$$

Thus, $Av \in V_\mu$, that is, V_μ is an invariant subspace of A. As a linear transformation on V_μ, A has an eigenvalue λ and a corresponding eigenvector v_1 in V_μ. Put in symbols,

$$Av_1 = \lambda v_1, \quad Bv_1 = \mu v_1, \quad v_1 \in V_\mu.$$

We may assume that v_1 is a unit vector. Extend v_1 to a unitary matrix U_1, that is, U_1 is a unitary matrix whose first column is v_1. By computation, we have

$$U_1^* A U_1 = \begin{pmatrix} \lambda & \alpha \\ 0 & C \end{pmatrix} \quad \text{and} \quad U_1^* B U_1 = \begin{pmatrix} \mu & \beta \\ 0 & D \end{pmatrix},$$

where $C, D \in \mathbb{M}_{n-1}$, and α and β are some row vectors.

It follows from $AB = BA$ that $CD = DC$. The induction hypothesis guarantees a unitary matrix $U_2 \in \mathbb{M}_{n-1}$ such that $U_2^* C U_2$ and $U_2^* D U_2$ are both upper-triangular. Let

$$U = U_1 \begin{pmatrix} 1 & 0 \\ 0 & U_2 \end{pmatrix}.$$

Then U, a product of two unitary matrices, is unitary, and U^*AU and U^*BU are both upper-triangular. ∎

Problems

1. Let A be a diagonal matrix with different diagonal entries. If B is a matrix such that $AB = BA$, show that B is also diagonal.

2. For square matrices A and B of the same size, if $AB = BA$ and if B has n distinct eigenvalues, show that A and B can have the same set of n linearly independent eigenvectors.

3. Let $f(x)$ be a polynomial and let A be an n-square matrix. Show that for any n-square invertible matrix P,
$$f(P^{-1}AP) = P^{-1}f(A)P$$
and that there exists a unitary matrix U such that both U^*AU and $U^*f(A)U$ are upper-triangular.

4. Show that the adjoints, inverses, sums, products, and polynomials of upper-triangular matrices are upper-triangular.

5. Show that every square matrix is a sum of two commuting matrices.

6. If A and B are two matrices such that $AB = I_m$ and $BA = I_n$, show that $m = n$, $AB = BA = I$, and $B = A^{-1}$.

7. Let A, B, and C be matrices such that $AB = CA$. Show that for any polynomial $f(x)$
$$Af(B) = f(C)A.$$

8. Is it true that any linear transformation on a vector space over $\mathbb{R}$ has at least a real eigenvalue?

9. Let A and B be commuting matrices. If A has k distinct eigenvalues, show that B has at least k linearly independent eigenvectors. Does it follow that B has k distinct eigenvalues?

10. Let A and B be n-square matrices. If $AB = BA$, what are the eigenvalues of $A + B$ and AB in terms of those of A and B?

11. Let $A, B \in \mathbb{M}_n$. If $AB = BA$, find the eigenvalues of the matrix
$$\begin{pmatrix} A & B \\ B & -A \end{pmatrix}.$$

12. What matrices in $\mathbb{M}_n$ commute with all diagonal matrices? With all Hermitian matrices? With all matrices in $\mathbb{M}_n$?

13. Let A and B be square complex matrices. If A commutes with B and B^*, show that $A + A^*$ commutes with $B + B^*$.

14. Show that Theorem 3.2 holds for more than two commuting matrices.

15. What conclusion can be drawn from Theorem 3.2 if B is assumed to be the identity matrix?

16. Let A and B be n- and m-square matrices, respectively, with $m \leq n$. If $AP = PB$ for an $n \times m$ matrix P with columns linearly independent, show that every eigenvalue of B is an eigenvalue of A.

17. If A and B are nonsingular matrices such that $AB - BA$ is singular, show that 1 is an eigenvalue of $A^{-1}B^{-1}AB$.

18. Let rank $(AB - BA) \leq 1$. If A or B has an eigenspace of dimension greater than or equal to 2, show that A and B have a common eigenvector. Find a common eigenvector for

$$A = \begin{pmatrix} 1 & -1 \\ 1 & -1 \end{pmatrix}, \quad B = \begin{pmatrix} 1 & 0 \\ 1 & 0 \end{pmatrix}.$$

(Note: In fact, rank $(AB - BA) \leq 1$ alone ensures the conclusion.)

19. Let A and B be $2n \times 2n$ matrices partitioned conformally as

$$A = \begin{pmatrix} A_{11} & 0 \\ 0 & A_{22} \end{pmatrix}, \quad B = \begin{pmatrix} B_{11} & B_{12} \\ B_{21} & B_{22} \end{pmatrix}.$$

If A commutes with B, show that A_{11} and A_{22} commute with B_{11} and B_{22}, respectively, and that for any polynomial $f(x)$

$$f(A_{11})B_{12} = B_{12}f(A_{22}), \quad f(A_{22})B_{21} = B_{21}f(A_{11}).$$

In particular, if $A_{11} = a_1 I$ and $A_{22} = a_2 I$ with $a_1 \neq a_2$, conclude that $B_{12} = B_{21} = 0$, and thus $B = B_{11} \oplus B_{22}$.

3.2 Matrix Decompositions

Factorizations of matrices into some special sorts of matrices via similarity are of fundamental importance in matrix theory. We will study the following decompositions of matrices in this section: the Schur decomposition, spectral decomposition, singular value decomposition, and polar decomposition. We will also continue our study of Jordan decomposition in later sections.

Theorem 3.3 (Schur Decomposition) *Let $\lambda_1, \lambda_2, \ldots, \lambda_n$ be the eigenvalues of $A \in \mathbb{M}_n$. Then there exists a unitary matrix $U \in \mathbb{M}_n$ such that U^*AU is an upper-triangular matrix. In symbols,*

$$U^*AU = \begin{pmatrix} \lambda_1 & & & * \\ & \lambda_2 & & \\ & & \ddots & \\ 0 & & & \lambda_n \end{pmatrix}.$$

PROOF. This theorem is a special result of Theorem 3.2. We present a pure matrix proof below without using the theory of vector spaces.

Let x_1 be a unit eigenvector of A belonging to eigenvalue λ_1:

$$Ax_1 = \lambda_1 x_1, \quad x_1 \neq 0.$$

Extend x_1 to a unitary matrix $S = (x_1, y_2, \ldots, y_n)$. Then

$$\begin{aligned} AS &= (Ax_1, Ay_2, \ldots, Ay_n) \\ &= (\lambda_1 x_1, Ay_2, \ldots, Ay_n) \\ &= S(u, S^{-1}Ay_2, \ldots, S^{-1}Ay_n), \end{aligned}$$

where $u = (\lambda_1, 0, \ldots, 0)^T$. Thus, we can write

$$S^*AS = \begin{pmatrix} \lambda_1 & v \\ 0 & B \end{pmatrix},$$

where v is a row vector and $B \in \mathbb{M}_{n-1}$.

Applying the induction hypothesis on B, we have a unitary matrix T of size $n-1$ such that T^*BT is upper-triangular. Let

$$U = S \begin{pmatrix} 1 & 0 \\ 0 & T \end{pmatrix}.$$

Then U, a product of two unitary matrices, is unitary, and U^*AU is upper-triangular. It is obvious that the diagonal entries λ_i of the upper-triangular matrix are the eigenvalues of A. ∎

A weaker statement is that of *triangularization*: For every $A \in \mathbb{M}_n$ there exists an invertible P such that $P^{-1}AP$ is upper-triangular.

Schur triangularization is one of the most important theorems in linear algebra and matrix theory. It will be used repeatedly in this book. As an application, we see by taking the conjugate transpose that any Hermitian matrix A (i.e., $A^* = A$) is unitarily diagonalizable. The same is true for normal matrices A, since the matrix identity $A^*A = AA^*$, together with the Schur decomposition of A, implies the desired decomposition form of A (Problem 4).

A *positive semidefinite matrix* $A \in \mathbb{M}_n$, by definition, $x^*Ax \geq 0$ for all $x \in \mathbb{C}^n$, has the similar structure. To see this, it suffices to show that a positive semidefinite matrix is necessarily Hermitian. This goes as follows:

Since $x^*Ax \geq 0$ for every $x \in \mathbb{C}^n$, we have, by taking x to be the column vector with the sth component 1, the tth component $c \in \mathbb{C}$, and 0 elsewhere,

$$x^*Ax = a_{ss} + a_{tt}|c|^2 + a_{ts}\bar{c} + a_{st}c \geq 0.$$

It follows that each diagonal entry a_{ss} is nonnegative by putting $c = 0$ and that $a_{st} = \overline{a_{ts}}$ or $A^* = A$ by putting $c = 1, i$, respectively.

It is immediate that the eigenvalues λ of a positive semidefinite A are nonnegative since $x^*Ax = \lambda x^*x$ for any eigenvector x of λ.

We summarize these discussions in the following theorem.

Theorem 3.4 (Spectral Decomposition) *Let A be an n-square complex matrix with eigenvalues $\lambda_1, \lambda_2, \ldots, \lambda_n$. Then A is normal if and only if A is unitarily diagonalizable, that is, there exists a unitary matrix U such that*

$$U^*AU = \operatorname{diag}(\lambda_1, \lambda_2, \ldots, \lambda_n).$$

In particular, A is Hermitian if and only if the λ_i are real and is positive semidefinite if and only if the λ_i are nonnegative.

As a result, by taking the square roots of the λ_i in the decomposition, we see that for any positive semidefinite matrix A, there exists a positive semidefinite matrix B such that $A = B^2$. Such a matrix B can be shown to be unique (Section 6.1) and is called a *square root* of A, denoted by $A^{\frac{1}{2}}$. In addition, we shall write

$$A \geq 0$$

if A is positive semidefinite, and $A > 0$ if $x^*Ax > 0$ for all $0 \neq x \in \mathbb{C}^n$.

The *singular values* of a matrix A are defined to be the nonnegative square roots of the eigenvalues of A^*A, which is positive semidefinite, for $x^*(A^*A)x = (Ax)^*(Ax) \geq 0$.

Theorem 3.5 (Singular Value Decomposition) *Let A be an $m \times n$ matrix with nonzero singular values $\sigma_1, \sigma_2, \ldots, \sigma_r$. Then there exist unitary matrices $U \in \mathbb{M}_m$ and $V \in \mathbb{M}_n$ such that*

$$A = U \begin{pmatrix} D & 0 \\ 0 & 0 \end{pmatrix} V, \qquad (3.1)$$

where $D = \mathrm{diag}(\sigma_1, \sigma_2, \ldots, \sigma_r)$. Thus, the rank of A is equal to r.

PROOF. If A is a number c, say, then the absolute value $|c|$ is the singular value of A, and $A = |c|e^{i\theta}$ for some $\theta \in \mathbb{R}$. If A is a nonzero row or column vector, say, $A = (a_1, \ldots, a_n)$, then σ_1 is the norm of the vector A. Let V be a unitary matrix with the first row the unit vector $(\frac{1}{\sigma_1}a_1, \ldots, \frac{1}{\sigma_1}a_n)$. Then $A = (\sigma_1, 0, \ldots, 0)V$.

We now assume $m > 1$, $n > 1$, and $A \neq 0$, and let u_1 be a unit eigenvector of A^*A belonging to σ_1^2, that is,

$$(A^*A)u_1 = \sigma_1^2 u_1, \quad u_1^* u_1 = 1.$$

Let

$$v_1 = \frac{1}{\sigma_1} A u_1.$$

Then v_1 is a unit vector and a simple computation gives $u_1^* A^* v_1 = \sigma_1$.

Let P and Q be unitary matrices with u_1 and v_1 as the first column, respectively. Then, with $A^*v_1 = \sigma_1 u_1$ and $(Au_1)^* = \sigma_1 v_1^*$,

$$P^*A^*Q = \begin{pmatrix} \sigma_1 & 0 \\ 0 & B \end{pmatrix} \quad \text{or} \quad A = Q \begin{pmatrix} \sigma_1 & 0 \\ 0 & B^* \end{pmatrix} P^*$$

for some $(n-1) \times (m-1)$ matrix B. The assertion follows by repeating the process on B^*. ∎

If A is an n-square matrix, then U and V are n-square unitary matrices, and $A = U(D \oplus 0)V$. By inserting U^*U between the block matrix and V in (3.1), we have

Theorem 3.6 (Polar Decomposition) *For any square matrix A, there exist a positive semidefinite P and a unitary U such that*

$$A = PU.$$

Problems

1. Let A be a square matrix partitioned as, with B, D square,

$$A = \begin{pmatrix} B & C \\ 0 & D \end{pmatrix}.$$

   Show that every eigenvalue of B or D is an eigenvalue of A.

2. Find a matrix P so that $P^{-1}AP$ is diagonal, then compute A^5, where

$$A = \begin{pmatrix} 2 & i \\ i & i \end{pmatrix} \quad \text{or} \quad \begin{pmatrix} 1 & 2 \\ 4 & 3 \end{pmatrix}.$$

3. Show that the complex symmetric A is not diagonalizable, where

$$A = \begin{pmatrix} 1 & i \\ i & -1 \end{pmatrix}.$$

   That is, $P^{-1}AP$ is not diagonal for any invertible matrix P.

4. If A is an upper-triangular matrix such that $A^*A = AA^*$, show that A is in fact a diagonal matrix.

5. Let A be an n-square complex matrix. Show that
$$x^*Ax = 0 \text{ for all } x \in \mathbb{C}^n \quad \Leftrightarrow \quad A = 0$$
and
$$x^T Ax = 0 \text{ for all } x \in \mathbb{R}^n \quad \Leftrightarrow \quad A^T = -A.$$

6. Let $A \in \mathbb{M}_n$. Show that if λ is an eigenvalue of A, then λ^k is an eigenvalue of A^k, and that $\alpha \in \mathbb{C}$ is an eigenvalue of $f(A)$ if and only if $\alpha = f(\lambda)$ for some eigenvalue λ of A, where f is a polynomial.

7. Show that if $A \in \mathbb{M}_n$ has n distinct eigenvalues, then A is diagonalizable. Does the converse hold?

8. Let A be an n-square positive semidefinite matrix. Show that
   (a) $X^*AX \geq 0$ for every $n \times m$ matrix X,
   (b) every principal submatrix of A is positive semidefinite.

9. Let
$$A = \begin{pmatrix} 0 & 1 & 0 \\ 0 & 0 & -1 \\ 0 & 0 & 0 \end{pmatrix}, \quad B = \begin{pmatrix} 0 & 0 & 0 \\ 1 & 0 & 0 \\ 0 & 1 & 0 \end{pmatrix}.$$
   (a) Show that $A^3 = B^3 = C^3 = 0$, where $C = \lambda A + \mu B$, $\lambda, \mu \in \mathbb{C}$.
   (b) Does there exist an integer k such that $(AB)^k = 0$?
   (c) Does there exist a nonsingular matrix P such that $P^{-1}AP$ and $P^{-1}BP$ are both upper-triangular?

10. Let A be an n-square complex matrix. Show that
    (a) (*QR Factorization*) there exist a unitary matrix Q and an upper-triangular matrix R such that $A = QR$. Further Q and R can be chosen to be real if A is real;
    (b) (*LU Factorization*) if all the leading principal minors of A are nonzero, then $A = LU$, where L and U are lower- and upper-triangular matrices, respectively.

11. Let $A \in \mathbb{M}_n$ have rank r. Show that A is normal if and only if
$$A = \sum_{i=1}^{r} \lambda_i u_i u_i^*,$$
where λ_i are complex numbers and u_i are column vectors of a unitary matrix. Further show that A is Hermitian if and only if all λ_i are real, and A is positive semidefinite if and only if all λ_i are nonnegative.

12. Let A be an $m \times n$ complex matrix with rank r. Show that
    (a) A has r nonzero singular values,
    (b) A has at most r nonzero eigenvalues,
    (c) $A = U_r D_r V_r$, where U_r is $m \times r$, $D_r = \operatorname{diag}(\sigma_1, \ldots, \sigma_r)$, V_r is $r \times n$, all of rank r, and $U_r^* U_r = V_r V_r^* = I_r$,
    (d) $A = \sum_{i=1}^{r} \sigma_i u_i v_i^*$, where the σ_i are the singular values of A, and u_i and v_i are column vectors of some unitary matrices.

13. Let A and B be upper-triangular matrices with positive diagonal entries. If $A = UB$ for some matrix U, show that $U = I$ and $A = B$.

14. Let A be a square matrix. Show that A has a zero singular value if and only if A has a zero eigenvalue. Does it follow that the number of zero singular values is equal to that of the zero eigenvalues?

15. Show that two Hermitian matrices are (unitarily) similar if and only if they have the same set of eigenvalues.

16. Let $A = P_1 U_1 = P_2 U_2$ be two polar decompositions of A. Show that $U_1 = U_2$ and $P_1 = P_2$ if A is nonsingular. What if A is singular?

17. Show that if A is an n-square complex matrix, then there exist nonsingular matrices P and Q such that $(PA)^2 = PA$, $(AQ)^2 = AQ$.

18. Let $A \in \mathbb{M}_n$. Show that $(A^*A)^{\frac{1}{2}} = UA$ for some unitary matrix U.

19. Show that $|\lambda_1 \cdots \lambda_n| = \sigma_1 \cdots \sigma_n$ for any $A \in \mathbb{M}_n$, where the λ_i and σ_i are the eigenvalues and singular values of A, respectively.

20. What can be said about $A \in \mathbb{M}_n$ if all its singular values are equal? Can the same conclusion be drawn in the case of eigenvalues?

21. If A is a matrix with eigenvalues λ_i and singular values σ_i, show that
$$\sum_i |\lambda_i|^2 \leq \operatorname{tr}(A^*A) = \operatorname{tr}(AA^*) = \sum_{i,j} |a_{ij}|^2 = \sum_i \sigma_i^2.$$

Equality occurs if and only if A is normal. Use this and the matrix with $(i, i+1)$-entry $\sqrt{x_i}$, $i = 1, 2, \ldots, n-1$, $(n,1)$-entry $\sqrt{x_n}$, and 0 elsewhere to show the arithmetic mean-geometric mean inequality
$$\left(\prod_{i=1}^{n} x_i \right)^{\frac{1}{n}} \leq \frac{1}{n} \sum_{i=1}^{n} x_i.$$

3.3 Annihilating Polynomials of Matrices

Given a polynomial in λ with complex coefficients $a_m, a_{m-1}, \ldots, a_0$,
$$f(\lambda) = a_m \lambda^m + a_{m-1} \lambda^{m-1} + \cdots + a_1 \lambda + a_0,$$
one can always define a matrix polynomial for $A \in \mathbb{M}_n$ by
$$f(A) = a_m A^m + a_{m-1} A^{m-1} + \cdots + a_1 A + a_0 I.$$
We consider in this section the annihilating polynomials of a matrix, that is, the polynomials $f(\lambda)$ for which $f(A) = 0$. Particular attention will be paid to the characteristic and minimal polynomials.

Theorem 3.7 *Let A be an n-square complex matrix. Then there exists a nonzero polynomial $f(\lambda)$ over $\mathbb{C}$ such that $f(A) = 0$.*

PROOF. $\mathbb{M}_n$ is a vector space over $\mathbb{C}$ of dimension n^2. Thus, any n^2+1 vectors in $\mathbb{M}_n$ are linearly dependent. In particular, the matrices
$$I,\ A,\ A^2,\ \ldots,\ A^{n^2}$$
are linearly dependent, namely, there exist numbers $a_0, a_1, a_2, \ldots, a_{n^2}$, not all zero, such that
$$a_0 I + a_1 A + a_2 A^2 + \cdots + a_{n^2} A^{n^2} = 0.$$
Set
$$f(\lambda) = a_0 + a_1 \lambda + a_2 \lambda^2 + \cdots + a_{n^2} \lambda^{n^2}.$$
Then $f(A) = 0$, as desired. ∎

Theorem 3.8 (Cayley-Hamilton) *Let A be an n-square complex matrix and let $p(\lambda)$ be the characteristic polynomial of A, that is,*
$$p(\lambda) = \det(\lambda I - A).$$
Then
$$p(A) = 0.$$

PROOF. Let the eigenvalues of A be $\lambda_1, \lambda_2, \ldots, \lambda_n$. We write A by triangularization as, for some invertible matrix P,

$$A = P^{-1}TP,$$

where T is an upper-triangular matrix with $\lambda_1, \lambda_2, \ldots, \lambda_n$ on the diagonal. Factor the characteristic polynomial $p(\lambda)$ of A as

$$p(\lambda) = (\lambda - \lambda_1)(\lambda - \lambda_2) \cdots (\lambda - \lambda_n).$$

Then
$$p(A) = p(P^{-1}TP) = P^{-1}p(T)P.$$

Note that
$$p(T) = (T - \lambda_1 I)(T - \lambda_2 I) \cdots (T - \lambda_n I).$$

It can be shown inductively that $(T - \lambda_1 I) \cdots (T - \lambda_k I)$ has the first k columns equal to zero, $1 \leq k \leq n$. Thus, $p(T) = 0$, and $p(A) = 0$. ∎

A monic polynomial $m(\lambda)$ is called the *minimal polynomial* of a matrix A if $m(A) = 0$ and it is of the smallest degree in the set

$$\{f(\lambda) : f(A) = 0\}.$$

It is immediate that if $f(A) = 0$, then $m(\lambda)$ divides $f(\lambda)$, or, in symbol, $m(\lambda)|f(\lambda)$, since otherwise we may write

$$f(\lambda) = q(\lambda)m(\lambda) + r(\lambda),$$

where $r(\lambda) \neq 0$ is of smaller degree than $m(\lambda)$ and $r(A) = 0$, a contradiction. In particular, the minimal polynomial divides the characteristic polynomial. Note that both the characteristic polynomial and the minimal polynomial are uniquely determined by its matrix.

Theorem 3.9 *Similar matrices have the same minimal polynomial.*

PROOF. Let A and B be similar matrices such that $A = P^{-1}BP$ for some nonsingular matrix P, and let $m_A(\lambda)$ and $m_B(\lambda)$ be the minimal polynomials of A and B, respectively. Then

$$m_B(A) = m_B(P^{-1}BP) = P^{-1}m_B(B)P = 0.$$

Thus, $m_A(\lambda)$ divides $m_B(\lambda)$. Similarly, $m_B(\lambda)$ divides $m_A(\lambda)$. Hence $m_A(\lambda) = m_B(\lambda)$ since they are both of leading coefficient 1. ∎

An effective method of computing minimal polynomials is given in the next section in conjunction with the discussion of Jordan canonical forms of square matrices.

Problems

1. Find the characteristic and minimal polynomials of the matrices
$$\begin{pmatrix} 0 & 1 \\ 0 & 0 \end{pmatrix}, \quad \begin{pmatrix} 1 & 1 \\ 0 & 1 \end{pmatrix}, \quad \begin{pmatrix} 1 & 1 \\ 1 & 1 \end{pmatrix}.$$

2. Find the characteristic and minimal polynomials of the matrices
$$\begin{pmatrix} 0 & 0 & c \\ 1 & 0 & b \\ 0 & 1 & a \end{pmatrix}, \quad \begin{pmatrix} \lambda & 1 & 0 \\ 0 & \lambda & 1 \\ 0 & 0 & \lambda \end{pmatrix}.$$

3. Find a nonzero polynomial $p(x)$ such that $p(A) = 0$, where
$$A = \begin{pmatrix} 2 & -2 & 0 \\ -2 & 1 & -2 \\ 0 & 2 & 0 \end{pmatrix}.$$

4. Compute $\det(AB - A)$ and $f(A)$, where $f(\lambda) = \det(\lambda B - A)$ and
$$A = \begin{pmatrix} 1 & 0 \\ 1 & 0 \end{pmatrix}, \quad B = \begin{pmatrix} 1 & 1 \\ 0 & 0 \end{pmatrix}.$$

5. Let $A \in \mathbb{M}_n$. Show that there exists a polynomial $f(x)$ with real coefficients such that $f(A) = 0$.

6. Let A and B be $n \times n$ matrices, and let $f(\lambda) = \det(\lambda I - B)$. Show that $f(A)$ is invertible if and only if A and B have no common eigenvalues.

7. Let A and B be square matrices of the same size. Show that if A and B are similar, then so are $f(A)$ and $f(B)$ for any polynomial f.

8. Let A and B be $n \times n$ matrices. If A and B are similar, show that $f(A) = 0$ if and only if $f(B) = 0$. Is the converse true?

9. Show that $\operatorname{rank}(AB) = n - 2$ if A and B are n-square upper-triangular matrices of rank $n - 1$ with diagonal entries zero.

SEC. 3.3 ANNIHILATING POLYNOMIALS OF MATRICES 73

10. Let $A_1, \ldots, A_m$ be upper-triangular matrices in $\mathbb{M}_n$. If they all have diagonal entries zero, show that $A_1 \cdots A_m = 0$ when $m \geq n$.

11. Explain what is wrong with the following proof of the Cayley-Hamilton theorem: Since $p(\lambda) = \det(\lambda I - A)$, plugging A for λ directly in both sides gives $p(A) = \det(A - A) = 0$.

12. As is known, for square matrices A and B of the same size, AB and BA have the same characteristic polynomial (Section 2.5). Do they have the same minimal polynomial?

13. Let $f(x)$ be a polynomial and let λ be an eigenvalue of a square matrix A. Show that if $f(A) = 0$, then $f(\lambda) = 0$.

14. Let $v \in \mathbb{C}^n$ and $A \in \mathbb{M}_n$. If $f(\lambda)$ is the monic polynomial with the smallest degree such that $f(A)v = 0$, show that $f(\lambda)$ divides $m_A(\lambda)$.

15. Let A and B be n-square complex matrices. Show that

$$AX - XB = 0 \quad \Rightarrow \quad f(A)X - Xf(B) = 0$$

for every polynomial f. In addition, if A and B have no common eigenvalues, then $AX - XB = 0$ has only the solution $X = 0$.

16. Let A and B be $2n \times 2n$ matrices partitioned conformally as

$$A = \begin{pmatrix} A_{11} & 0 \\ 0 & A_{22} \end{pmatrix}, \quad B = \begin{pmatrix} B_{11} & B_{12} \\ B_{21} & B_{22} \end{pmatrix}.$$

If $AB = BA$ and A_{11} and A_{22} have no common eigenvalue, show that

$$B_{12} = B_{21} = 0.$$

17. Show that for any nonsingular matrix A, matrices A^{-1} and $\mathrm{adj}(A)$ can be expressed as polynomials in A.

18. Express J^{-1} as a polynomial in J, where

$$J = \begin{pmatrix} 1 & 1 & & 0 \\ & 1 & \ddots & \\ & & \ddots & 1 \\ 0 & & & 1 \end{pmatrix}.$$

3.4 Jordan Canonical Forms

We saw in Section 3.2 that a square matrix is similar (and even unitarily similar) to an upper-triangular matrix. We now discuss the upper-triangular matrices and give simpler structures.

The main theorem of this section is the *Jordan decomposition*, which states that every square complex matrix is similar (not necessarily unitarily similar) to a direct sum of Jordan blocks, referred to as *Jordan canonical form* or simply *Jordan form*. A *Jordan block* is a square matrix in form

$$J_x = \begin{pmatrix} x & 1 & & 0 \\ & x & \ddots & \\ & & \ddots & 1 \\ 0 & & & x \end{pmatrix}. \qquad (3.2)$$

For this purpose we will introduce λ-matrices and make use of elementary operations that bring λ-matrices to standard form. We then show that two matrices A and B in $\mathbb{M}_n$ are similar if and only if their λ-matrices $\lambda I - A$ and $\lambda I - B$ can be brought to the same standard form. Thus, a square matrix A is similar to its Jordan form that is determined by the standard form of the λ-matrix $\lambda I - A$.

To proceed, a λ-*matrix* is a matrix whose entries are complex polynomials in λ. For instance,

$$\begin{pmatrix} 1 & \lambda^2 - \sqrt{2} & (\lambda+1)^2 \\ \frac{1}{2}\lambda - 1 & 0 & 1 - 2\lambda - \lambda^2 \\ -1 & 1 & \lambda - i \end{pmatrix}$$

is a λ-matrix, for every entry is a polynomial in λ (or a constant). Note that the minimal polynomial of J_x of size t, say, in (3.2) is

$$m(\lambda) = (\lambda - x)^t.$$

Elementary operations on λ-matrices are similar to those on numerical matrices. (Note that division by a nonzero polynomial is

SEC. 3.4   JORDAN CANONICAL FORMS   75

not permitted.) *Elementary λ-matrices* and *invertible λ-matrices* are similarly defined as those of numerical matrices.

Any square numerical matrix can be brought into a diagonal matrix with 1 and 0 on the main diagonal by elementary operations. Likewise, λ-matrices can be brought into *standard form*

$$\begin{pmatrix} d_1(\lambda) & & & & & 0 \\ & \ddots & & & & \\ & & d_k(\lambda) & & & \\ & & & 0 & & \\ & & & & \ddots & \\ 0 & & & & & 0 \end{pmatrix}, \qquad (3.3)$$

where $d_i(\lambda)|d_{i+1}(\lambda)$, $i = 1, \ldots, k-1$, and each $d_i(\lambda)$ is 1 or monic. Therefore, for any λ-matrix $A(\lambda)$ there exist elementary λ-matrices $P_s(\lambda), \ldots, P_1(\lambda)$ and $Q_1(\lambda), \ldots, Q_t(\lambda)$ such that

$$P_s(\lambda) \cdots P_1(\lambda) A(\lambda) Q_1(\lambda) \cdots Q_t(\lambda) = D(\lambda)$$

is in the standard form (3.3).

If $A(\lambda)$ is an invertible λ-matrix, that is, $B(\lambda)A(\lambda) = I$ for some λ-matrix $B(\lambda)$ of the same size, then, by taking determinants, we see that $\det A(\lambda)$ is a nonzero constant. Conversely, if $\det A(\lambda)$ is a nonzero constant, then $(\det A(\lambda))^{-1} \operatorname{adj}(A(\lambda))$ is also a λ-matrix and it is the inverse of $A(\lambda)$. Moreover, a square λ-matrix is invertible if and only if its standard form (3.3) is the identity matrix and if and only if it is a product of elementary λ-matrices.

For the λ-matrix $\lambda I - A$, $A \in \mathbb{M}_n$, $k = n$. The $d_i(\lambda)$ are called the *invariant factors* of A, and the divisors of $d_i(\lambda)$ factored into the form $(\lambda - x)^t$ for some constant x and positive integer t are called the *elementary divisors* of A. To illustrate this, look at the example

$$A = \begin{pmatrix} -1 & 0 & 0 \\ 1 & 1 & 2 \\ 3 & 0 & 1 \end{pmatrix}.$$

We perform elementary operations on the λ-matrix

$$\lambda I - A = \begin{pmatrix} \lambda+1 & 0 & 0 \\ -1 & \lambda-1 & -2 \\ -3 & 0 & \lambda-1 \end{pmatrix}.$$

Interchange row 1 and row 2 times -1 to get a 1 for the (1, 1)-position:
$$\begin{pmatrix} 1 & 1-\lambda & 2 \\ \lambda+1 & 0 & 0 \\ -3 & 0 & \lambda-1 \end{pmatrix}.$$

Add row 1 times $-(\lambda+1)$ and 3 to rows 2 and 3, respectively, to get 0 below 1:
$$\begin{pmatrix} 1 & 1-\lambda & 2 \\ 0 & (\lambda-1)(\lambda+1) & -2(\lambda+1) \\ 0 & -3(\lambda-1) & \lambda+5 \end{pmatrix}.$$

Add row 3 times 2 to row 2 to get a nonzero number 8:
$$\begin{pmatrix} 1 & 1-\lambda & 2 \\ 0 & \lambda^2-6\lambda+5 & 8 \\ 0 & -3(\lambda-1) & \lambda+5 \end{pmatrix}.$$

Interchange column 2 and column 3 to get a nonzero number for the (2, 2)-position:
$$\begin{pmatrix} 1 & 2 & 1-\lambda \\ 0 & 8 & \lambda^2-6\lambda+5 \\ 0 & \lambda+5 & -3(\lambda-1) \end{pmatrix}.$$

Subtract the second row times $\frac{1}{8}(\lambda+5)$ from row 3 to get
$$\begin{pmatrix} 1 & 2 & 1-\lambda \\ 0 & 8 & \lambda^2-6\lambda+5 \\ 0 & 0 & -\frac{1}{8}(\lambda-1)^2(\lambda+1) \end{pmatrix},$$

which gives the standard form at once (by column operations)
$$\begin{pmatrix} 1 & 0 & 0 \\ 0 & 1 & 0 \\ 0 & 0 & (\lambda+1)(\lambda-1)^2 \end{pmatrix}.$$

Thus, the invariant factors of A are
$$d_1(\lambda)=1, \quad d_2(\lambda)=1, \quad d_3(\lambda)=(\lambda+1)(\lambda-1)^2,$$

SEC. 3.4   JORDAN CANONICAL FORMS   77

and the elementary divisors of A are

$$\lambda + 1, \quad (\lambda - 1)^2.$$

Note that the matrix, a direct sum of two Jordan blocks,

$$J = \begin{pmatrix} -1 & 0 & 0 \\ 0 & 1 & 1 \\ 0 & 0 & 1 \end{pmatrix} = (-1) \oplus \begin{pmatrix} 1 & 1 \\ 0 & 1 \end{pmatrix}$$

has the same invariant factors and elementary divisors as A.

In general, each elementary divisor $(\lambda - x)^t$ corresponds to a Jordan block in the form (3.2). Consider all the elementary divisors of a matrix A, find all the corresponding Jordan blocks, and form a direct sum of them. A profound conclusion is that A is similar to this direct sum. To this end, we need to show a fundamental theorem of λ-matrix theory.

Theorem 3.10 *Let A and B be n-square complex matrices. Then A and B are similar if and only if $\lambda I - A$ and $\lambda I - B$ have the same standard form. Equivalently, there exist λ-matrices $P(\lambda)$ and $Q(\lambda)$ that are products of elementary λ-matrices such that*

$$P(\lambda)(\lambda I - A)Q(\lambda) = \lambda I - B.$$

This will imply our main theorem on the Jordan canonical form, which is one of the most useful results in linear algebra and matrix theory. The theorem itself is much more important than its proof. We sketch the proof of Theorem 3.10 as follows.

PROOF OUTLINE. If A and B are similar, then there exists an invertible complex matrix P such that $PAP^{-1} = B$. It follows that

$$P(\lambda I - A)P^{-1} = \lambda I - B.$$

To show the other way, let $P(\lambda)$ and $Q(\lambda)$ be invertible λ-matrices such that (we put $Q(\lambda)$ for $Q(\lambda)^{-1}$ on the right for convenience)

$$P(\lambda)(\lambda I - A) = (\lambda I - B)Q(\lambda). \tag{3.4}$$

Write (Problem 4)

$$P(\lambda) = (\lambda I - B)P_1(\lambda) + P, \quad Q(\lambda) = Q_1(\lambda)(\lambda I - A) + Q,$$

where $P_1(\lambda)$ and $Q_1(\lambda)$ are λ-matrices, P and Q numerical matrices.

Identity (3.4) implies $P_1(\lambda) - Q_1(\lambda) = 0$ by considering the degree of $P_1(\lambda) - Q_1(\lambda)$. It follows that $Q = P$, and thus $PA = BP$.

It remains to show that P is invertible. Assume $R(\lambda)$ is the inverse of $P(\lambda)$, or $P(\lambda)R(\lambda) = I$. Write $R(\lambda) = (\lambda I - A)R_1(\lambda) + R$, where R is a numerical matrix. With $PA = BP$, $I = P(\lambda)R(\lambda)$ gives

$$I = (\lambda I - B)T(\lambda) + PR, \qquad (3.5)$$

where

$$T(\lambda) = P_1(\lambda)(\lambda I - A)R_1(\lambda) + P_1(\lambda)R + PR_1(\lambda).$$

By considering the degree of both sides of (3.5), $T(\lambda)$ must be zero. Therefore, $I = PR$ and hence P is nonsingular. ∎

We thus have, based on the earlier discussions, the main result:

Theorem 3.11 (Jordan Decomposition) *Let A be a square complex matrix. Then there exists an invertible matrix P such that*

$$P^{-1}AP = J_1 \oplus \cdots \oplus J_s,$$

where the J_i are the Jordan blocks of A with the eigenvalues of A on the diagonal. The Jordan blocks are uniquely determined by A.

The uniqueness of the Jordan decomposition of A up to permutations of the diagonal Jordan blocks follows from the uniqueness of the standard form (3.3) of $\lambda I - A$. Two different sets of Jordan blocks will result in two different standard forms (3.3).

To find the minimal polynomial of a given matrix $A \in \mathbb{M}_n$, reduce $\lambda I - A$ by elementary operations to a standard form with invariant factors $d_1(\lambda), \ldots, d_n(\lambda)$, $d_i(\lambda)|d_{i+1}(\lambda)$, for $i = 1, \ldots, n-1$. Note that similar matrices have the same minimal polynomial (Theorem 3.9). Thus, $d_n(\lambda)$ is the minimal polynomial of A, since it is the minimal polynomial of the Jordan canonical form of A (Problem 12).

Theorem 3.12 *Let $p(\lambda)$ and $m(\lambda)$ be, respectively, the characteristic and minimal polynomials of matrix $A \in \mathbb{M}_n$. Let $d_1(\lambda), \ldots, d_n(\lambda)$ be the invariant factors of A, where $d_i(\lambda)|d_{i+1}$, $i = 1, \ldots, n-1$. Then*

$$p(\lambda) = d_1(\lambda) \cdots d_n(\lambda), \quad m(\lambda) = d_n(\lambda).$$

In the earlier example preceding Theorem 3.10, the characteristic and minimal polynomials of A are the same, and they are equal to

$$p(\lambda) = m(\lambda) = (\lambda + 1)(\lambda - 1)^2 = \lambda^3 - \lambda^2 - \lambda + 1.$$

Problems

1. Find the invariant factors, elementary divisors, characteristic and minimal polynomials, and the Jordan canonical form of the matrix

$$A = \begin{pmatrix} 3 & 1 & -3 \\ -7 & -2 & 9 \\ -2 & -1 & 4 \end{pmatrix}.$$

2. Show that A and B are similar but not unitarily similar, where

$$A = \begin{pmatrix} 0 & 2 \\ 0 & 0 \end{pmatrix}, \quad B = \begin{pmatrix} 0 & 3 \\ 0 & 0 \end{pmatrix}.$$

What is the Jordan canonical form J of A and B? Can one find an invertible matrix P such that

$$P^{-1}AP = P^{-1}BP = J?$$

3. Are the following matrices similar? Why?

$$\begin{pmatrix} 1 & 1 & 0 \\ 0 & 1 & 1 \\ 0 & 0 & 1 \end{pmatrix}, \quad \begin{pmatrix} 1 & 0 & 0 \\ 2 & 1 & 0 \\ 0 & 2 & 1 \end{pmatrix}, \quad \begin{pmatrix} 1 & 2 & 0 \\ 2 & 1 & 0 \\ 0 & 0 & 1 \end{pmatrix}.$$

4. Let $A \in \mathbb{M}_n$. If $P(\lambda)$ is an n-square λ-matrix, show that there exist a λ-matrix $S(\lambda)$ and a numerical matrix T such that

$$P(\lambda) = (\lambda I - A)S(\lambda) + T.$$

(Hint: Write $P(\lambda) = \lambda^m P_m + \lambda^{m-1} P_{m-1} + \cdots + \lambda P_1 + P_0$.)

5. Show that a λ-matrix is invertible if and only if it is a product of elementary λ-matrices.

6. Show that two matrices are similar if and only if they have the same set of Jordan blocks, counting the repeated ones.

7. Find the invariant factors, elementary divisors, characteristic and minimal polynomials for each of the following matrices:

$$\begin{pmatrix} 1 & 0 & 0 \\ 0 & 1 & 1 \\ 0 & 0 & 1 \end{pmatrix}, \quad \begin{pmatrix} -1 & 0 & 0 \\ 0 & 1 & 1 \\ 0 & 0 & 1 \end{pmatrix}, \quad \begin{pmatrix} 0 & 1 & 0 \\ 0 & 0 & 1 \\ 1 & 0 & 0 \end{pmatrix}, \quad \begin{pmatrix} 1 & 1 & 0 \\ 0 & 1 & 0 \\ 0 & 1 & 1 \end{pmatrix},$$

$$\begin{pmatrix} 1 & 1 & 0 & 0 \\ 0 & 1 & 0 & 0 \\ 0 & 0 & 1 & 1 \\ 0 & 0 & 0 & 1 \end{pmatrix}, \quad \begin{pmatrix} 1 & -1 & 0 & 0 \\ 0 & 1 & -1 & 0 \\ 0 & 0 & 1 & -1 \\ 0 & 0 & 0 & 1 \end{pmatrix}, \quad \begin{pmatrix} \lambda & 0 & 1 & 0 \\ 0 & \lambda & 0 & 1 \\ 0 & 0 & \lambda & 0 \\ 0 & 0 & 0 & \lambda \end{pmatrix}.$$

8. Find the Jordan canonical form of the matrix

$$P = \begin{pmatrix} 0 & 1 & 0 & \cdots & 0 \\ 0 & 0 & 1 & \cdots & 0 \\ \vdots & \vdots & \vdots & \vdots & \vdots \\ 0 & 0 & 0 & \cdots & 1 \\ 1 & 0 & 0 & \cdots & 0 \end{pmatrix}.$$

9. Let A be a square complex matrix with invariant factors

$$1, \quad \lambda(\lambda - 2), \quad \lambda^3(\lambda - 2).$$

Answer the following questions:

(a) What is the characteristic polynomial of A?

(b) What is the minimal polynomial of A?

(c) What are the elementary divisors of A?

(d) What is the size of A?

(e) What is the rank of A?

(f) What is the trace of A?

(g) What is the Jordan form of A?

10. Let J be a Jordan block. Find the Jordan forms of J^{-1} (if it exists) and J^2.

11. Show that every Jordan block J is similar to J^T via S:

$$S^{-1}JS = J^T,$$

where S is the backward identity matrix, that is, $s_{i,n-i+1} = 1$ for $i = 1, 2, \ldots, n$, and 0 elsewhere.

12. Show that the last invariant factor $d_n(\lambda)$ in the standard form of $\lambda I - A$ is the minimal polynomial of $A \in \mathbb{M}_n$.

13. If J is a Jordan block such that $J^2 = J$, show that $J = I$ or 0.

14. Let A be an $n \times n$ matrix. Show that $\operatorname{rank}(A^2) = \operatorname{rank}(A)$ implies $\operatorname{rank}(A^k) = \operatorname{rank}(A)$ for any integer $k > 0$ and $\mathbb{C}^n = \operatorname{Im} A \oplus \operatorname{Ker} A$.

15. Let A be an $n \times n$ matrix. Show that $\operatorname{rank}(A^k) = \operatorname{rank}(A^{k+1})$ for some positive integer $k \leq n$ and that $\operatorname{rank}(A^k) = \operatorname{rank}(A^m)$ for all positive integer $m > k$. In particular, $\operatorname{rank}(A^n) = \operatorname{rank}(A^{n+1})$.

16. Show that every matrix $A \in \mathbb{M}_n$ can be written as $A = B + C$, where $C^k = 0$ for some integer k, B is diagonalizable, and $BC = CB$.

17. Let A be a square complex matrix. If $Ax = 0$ whenever $A^2 x = 0$, show that A does not have any Jordan block of order more than 1 corresponding to eigenvalue 0.

18. Show that if matrix A has all eigenvalues equal to 1, then A^k is similar to A for every positive integer k. Discuss the converse.

19. Show that the dimension of the vector space of all the polynomials in A is equal to the degree of the minimal polynomial of A.

20. Let A be an $n \times n$ matrix such that $A^k v = 0$ and $A^{k-1} v \neq 0$ for some vector v and positive integer k. Show that $v, Av, \ldots, A^{k-1} v$ are linearly independent. What is the Jordan form of A?

21. Let $A \in \mathbb{M}_n$ be a Jordan block. Show that there exists a vector v such that $v, Av, \ldots, A^{n-1} v$ constitute a basis for $\mathbb{C}^n$.

22. If the eigenvalues of $A \in \mathbb{M}_n$ are all distinct, show that $v, Av, \ldots, A^{n-1} v$ are linearly independent for any $v \in \mathbb{C}^n$ with all $v_i \neq 0$.

23. Show that the characteristic polynomial coincides with the minimal polynomial for $A \in \mathbb{M}_n$ if and only if $v, Av, \ldots, A^{n-1} v$ are linearly independent for some vector $v \in \mathbb{C}^n$. What can be said about the Jordan form (or Jordan blocks) of A?

24. Let A be an n-square complex matrix. Show that for any nonzero vector $v \in \mathbb{C}^n$, there exists an eigenvector u of A that is contained in the span of $v, Av, A^2v, \dots$. (Hint: $v, Av, A^2v, \dots, A^k$ are linearly dependent for some k. Find a related polynomial then factor it out.)

25. Let A be an n-square complex matrix with the characteristic polynomial factored over the complex field $\mathbb{C}$ as

$$\det(\lambda I - A) = (\lambda - \lambda_1)^{r_1}(\lambda - \lambda_2)^{r_2}\cdots(\lambda - \lambda_s)^{r_s},$$

where $\lambda_1, \lambda_2, \dots, \lambda_s$ are the distinct eigenvalues of A. Show that following statements are equivalent:

   (a) A is diagonalizable, namely, A is similar to a diagonal matrix;
   (b) A has n linearly independent eigenvectors;
   (c) all the elementary divisors of $\lambda I - A$ are linear;
   (d) the minimal polynomial of A has no repeated zeros;
   (e) $\operatorname{rank}(\lambda I - A) = \operatorname{rank}(\lambda I - A)^2$ for every eigenvalue λ;
   (f) $\operatorname{rank}(cI - A) = \operatorname{rank}(cI - A)^2$ for every complex number c;
   (g) $(\lambda I - A)x = 0$ and $(\lambda I - A)^2 x = 0$ have the same solution space for every eigenvalue λ;
   (h) $(cI - A)x = 0$ and $(cI - A)^2 x = 0$ have the same solution space for every complex number c;
   (i) $\dim V_{\lambda_i} = r_i$ for each eigenspace V_{λ_i} of eigenvalue λ_i;
   (j) $\operatorname{rank}(\lambda_i I - A) = n - r_i$ for every eigenvalue λ_i;
   (k) $\operatorname{Im}(\lambda I - A) \cap \operatorname{Ker}(\lambda I - A) = \{0\}$ for every eigenvalue λ;
   (l) $\operatorname{Im}(cI - A) \cap \operatorname{Ker}(cI - A) = \{0\}$ for every complex number c.

26. Let $\mathcal{A}$ be a linear transformation on a finite-dimensional vector space. Let λ be an eigenvalue of $\mathcal{A}$. Show that each subspace $\operatorname{Ker}(\lambda \mathcal{I} - \mathcal{A})^k$, where k is a positive integer, is invariant under $\mathcal{A}$, and that

$$\operatorname{Ker}(\lambda \mathcal{I} - \mathcal{A}) \subseteq \operatorname{Ker}(\lambda \mathcal{I} - \mathcal{A})^2 \subseteq \operatorname{Ker}(\lambda \mathcal{I} - \mathcal{A})^3 \subseteq \cdots.$$

Conclude that for some positive integer m,

$$\operatorname{Ker}(\lambda \mathcal{I} - \mathcal{A})^m = \operatorname{Ker}(\lambda \mathcal{I} - \mathcal{A})^{m+1} = \cdots$$

and that

$$\bigcup_{k=1}^{\infty} \operatorname{Ker}(\lambda \mathcal{I} - \mathcal{A})^k = \operatorname{Ker}(\lambda \mathcal{I} - \mathcal{A})^m.$$

he Matrices A^T, $\overline{A}$, A^*, A^TA, A^*A, and $\overline{A}A$

·ices associated to a matrix A and often encountered are

$$A^T, \quad \overline{A}, \quad A^*, \quad A^TA, \quad A^*A, \quad \overline{A}A,$$

hen the inverse exists, where T, $-$, and $*$ represent trans-
gate, and transpose conjugate, respectively.
se matrices, except A^TA, are equal in rank:

) $= \operatorname{rank}(A^T) = \operatorname{rank}(\overline{A}) = \operatorname{rank}(A^*) = \operatorname{rank}(A^*A)$.

ntity is due to the fact that the equation systems

$$(A^*A)x = 0 \quad \text{and} \quad Ax = 0$$

e solution space.

13 *Let A be an n-square complex matrix. Then*

ilar to its transpose A^T,

ilar to A^* (equivalently $\overline{A}$) if and only if the Jordan
the nonreal eigenvalues of A occur in conjugate pairs,

imilar to AA^*,

nilar to $A\overline{A}$.

, recall from Theorem 3.10 that two matrices X and
and only if $\lambda I - X$ and $\lambda I - Y$ have the same standard
ous that the matrices $\lambda I - A$ and $\lambda I - A^T$ have the
form. Thus, A and A^T are similar. An alternative
is to verify that for every Jordan block J,

$$SJS^{-1} = J^T,$$

ackward identity matrix (Problem 11, Section 3.4).

For (2), let $J_1, \ldots, J_k$ be the Jordan blocks of A and let

$$P^{-1}AP = J_1 \oplus \cdots \oplus J_k$$

for some invertible matrix P. Taking the transpose conjugate gives

$$P^*A^*(P^*)^{-1} = J_1^* \oplus \cdots \oplus J_k^*.$$

The right-hand side, by (1), is similar to

$$\overline{J_1} \oplus \cdots \oplus \overline{J_k}.$$

Thus, if A and A^* are similar, then

$$J_1 \oplus \cdots \oplus J_k \quad \text{and} \quad \overline{J_1} \oplus \cdots \oplus \overline{J_k}$$

are similar. It follows by the uniqueness of Jordan decomposit that the Jordan blocks of nonreal complex eigenvalues of A m occur in conjugate pairs (Problem 6, Section 3.4).

For sufficiency, we may consider the special case

$$A = J \oplus \overline{J} \oplus R, \qquad ($$

where J and R are Jordan blocks, J is complex, and R is real. T

$$A^* = \overline{J}^T \oplus J^T \oplus R^T,$$

which is, by permutation, similar to

$$J^T \oplus \overline{J}^T \oplus R^T.$$

Using (1), (3.6) and (3.7) give the similarity of A^* and A.

(3) is by a singular value decomposition of A.

We have left to show (4) that $\overline{A}A$ is similar to $A\overline{A}$. It suffi show that $A\overline{A}$ and $\overline{A}A$ have the same Jordan decomposition (bl

The matrix identity

$$\begin{pmatrix} I & -A \\ 0 & I \end{pmatrix} \begin{pmatrix} A\overline{A} & 0 \\ \overline{A} & 0 \end{pmatrix} \begin{pmatrix} I & A \\ 0 & I \end{pmatrix} = \begin{pmatrix} 0 & 0 \\ \overline{A} & \overline{A}A \end{pmatrix}$$

gives the similarity of the block matrices

$$\begin{pmatrix} A\overline{A} & 0 \\ \overline{A} & 0 \end{pmatrix} \quad \text{and} \quad \begin{pmatrix} 0 & 0 \\ A & \overline{A}A \end{pmatrix}.$$

Thus, the nonsingular Jordan blocks of $A\overline{A}$ and $\overline{A}A$ are identical. (In general, this is true for AB and BA. See Problem 12). On the other hand, the singular Jordan blocks of $A\overline{A}$ and $\overline{A}A = \overline{A\overline{A}}$ are obviously the same. This concludes that $A\overline{A}$ and $\overline{A}A$ are similar. ∎

Following the discussion of the case where A is similar to A^* in the proof, one may obtain a more profound result. Consider the matrix with Jordan blocks of conjugate pairs

$$\begin{pmatrix} \lambda & 1 & 0 & 0 \\ 0 & \lambda & 0 & 0 \\ 0 & 0 & \overline{\lambda} & 1 \\ 0 & 0 & 0 & \overline{\lambda} \end{pmatrix},$$

which is similar via permutation to

$$\begin{pmatrix} \lambda & 0 & 1 & 0 \\ 0 & \overline{\lambda} & 0 & 1 \\ 0 & 0 & \lambda & 0 \\ 0 & 0 & 0 & \overline{\lambda} \end{pmatrix} = \begin{pmatrix} C(\lambda) & I \\ 0 & C(\lambda) \end{pmatrix},$$

where

$$C(\lambda) = \begin{pmatrix} \lambda & 0 \\ 0 & \overline{\lambda} \end{pmatrix}.$$

If $\lambda = a + bi$ with $a, b \in \mathbb{R}$, then we have by computation

$$\begin{pmatrix} -i & -i \\ 1 & -1 \end{pmatrix} \begin{pmatrix} \lambda & 0 \\ 0 & \overline{\lambda} \end{pmatrix} \begin{pmatrix} -i & -i \\ 1 & -1 \end{pmatrix}^{-1} = \begin{pmatrix} a & b \\ -b & a \end{pmatrix}.$$

Thus, matrices

$$\begin{pmatrix} \lambda & 0 & 1 & 0 \\ 0 & \overline{\lambda} & 0 & 1 \\ 0 & 0 & \lambda & 0 \\ 0 & 0 & 0 & \overline{\lambda} \end{pmatrix} \quad \text{and} \quad \begin{pmatrix} a & b & 1 & 0 \\ -b & a & 0 & 1 \\ 0 & 0 & a & b \\ 0 & 0 & -b & a \end{pmatrix}$$

are similar. These observations lead to the following theorem.

Theorem 3.14 *A square matrix A is similar to A^* (equivalently $\overline{A}$) if and only if A is similar to a real matrix.*

As a result, $\overline{A}A$ is similar to a real matrix. The ideas of pairing Jordan blocks are often used in the similarity theory of matrices.

Problems

1. Let A be an $m \times n$ matrix. Prove or disprove
   (a) $\operatorname{rank}(A) = \operatorname{rank}(A^T A)$,
   (b) $\operatorname{rank}(A^* A) = \operatorname{rank}(AA^*)$,
   (c) $A^T A$ is similar to AA^T.

2. Is it possible that $\overline{A}A = 0$ or $A^*A = 0$ for a nonzero $A \in \mathbb{M}_n$?

3. Let
$$A = \begin{pmatrix} 1 & 1 \\ 1 & 1 \end{pmatrix}, \quad B = \begin{pmatrix} 1 & 1 \\ -1 & -1 \end{pmatrix}.$$
   Compute AB and BA. Find $\operatorname{rank}(AB)$ and $\operatorname{rank}(BA)$. What are the Jordan forms of AB and BA? Are AB and BA similar?

4. If the nonreal eigenvalues of a square matrix A occur in conjugate pairs, does it follow that A is similar to A^*?

5. Show that the characteristic polynomial of $A\overline{A}$ has only real coefficients. Conclude that the nonreal eigenvalues of $A\overline{A}$ must occur in conjugate pairs.

6. Let $A = B + iC \in \mathbb{M}_n$ be nonsingular, where B and C are real square matrices. Show that $\overline{A} = A^{-1}$ if and only if $BC = CB$ and $B^2 + C^2 = I$. Find the conditions on B and C if $A^T = \overline{A} = A^{-1}$.

7. Let $A \in \mathbb{M}_n$. Show that the following statements are equivalent:
   (a) A is similar to A^* (equivalently $\overline{A}$);
   (b) the elementary divisors occur in conjugate pairs;
   (c) the invariant factors of A are all real coefficients.

   Are they equivalent to the statement "$\det(\lambda I - A)$ is real coefficient"?

8. If $A \in \mathbb{M}_n$ has only real eigenvalues, show that A is similar to A^*.

9. Let A be a nonsingular matrix. When is A^{-1} similar to A?

10. Let $A \in \mathbb{M}_n$ and B be $m \times n$. Let M be the $(m+n)$-square matrix
$$M = \begin{pmatrix} A & 0 \\ B & 0 \end{pmatrix}.$$
Show that the nonsingular Jordan blocks of A and M are identical.

11. Let A, B, C, and D be n-square complex matrices. If the matrices
$$\begin{pmatrix} A & B \\ 0 & 0 \end{pmatrix} \quad \text{and} \quad \begin{pmatrix} C & D \\ 0 & 0 \end{pmatrix}$$
are similar, does it follow that A and C are similar?

12. Let A and B be $m \times n$ complex matrices. Show that the nonsingular Jordan blocks of AB and BA are identical. Conclude that AB and BA have the same nonzero eigenvalues, including multiplicity.

13. If A and $B \in \mathbb{M}_n$ have no common eigenvalues, show that the following two block matrices are similar for any $X \in \mathbb{M}_n$:
$$\begin{pmatrix} A & X \\ 0 & B \end{pmatrix}, \quad \begin{pmatrix} A & 0 \\ 0 & B \end{pmatrix}.$$

14. Let A, B, and C be matrices of appropriate sizes. Show that
$$AX - YB = C$$
for some matrices X and Y if and only if the block matrices
$$\begin{pmatrix} A & C \\ 0 & B \end{pmatrix} \quad \text{and} \quad \begin{pmatrix} A & 0 \\ 0 & B \end{pmatrix}$$
are similar.

15. Let A and B be n-square complex matrices and let
$$M = \begin{pmatrix} A & B \\ -\overline{B} & \overline{A} \end{pmatrix}.$$
Show that

(a) the characteristic polynomial of M is of real coefficients,

(b) the eigenvalues of M occur in conjugate pairs with eigenvectors in forms $\begin{pmatrix} x \\ y \end{pmatrix} \in \mathbb{C}^{2n}$ and $\begin{pmatrix} -\overline{y} \\ \overline{x} \end{pmatrix} \in \mathbb{C}^{2n}$,

(c) the eigenvectors in (b) are linearly independent,

(d) $\det M \geq 0$. In particular, $\det(I + A\overline{A}) \geq 0$ for any $A \in \mathbb{M}_n$.

3.6 Numerical Range

The *numerical range*, also known as the *field of values*, of an n-square complex matrix A is defined to be

$$W(A) = \{x^* A x : \|x\| = 1, \ x \in \mathbb{C}^n\}.$$

For example, if

$$A = \begin{pmatrix} 1 & 0 \\ 0 & 0 \end{pmatrix},$$

then $W(A)$ is the closed interval $[0, 1]$, and if

$$A = \begin{pmatrix} 0 & 0 \\ 1 & 1 \end{pmatrix},$$

then $W(A)$ is the closed elliptical disc with foci at $(0,0)$ and $(1,0)$, minor axis 1, and major axis $\sqrt{2}$.

One of the celebrated and fundamental results on numerical range is the Toeplitz-Hausdorff convexity theorem.

Theorem 3.15 (Toeplitz-Hausdorff) *The numerical range of a square matrix is a convex compact subset of the complex plane.*

PROOF. For convexity, if $W(A)$ is a singleton, there is nothing to show. Suppose $W(A)$ has more than one point. We prove that the line segment joining any two distinct points in $W(A)$ lies in $W(A)$, that is, if $u, v \in W(A)$, then $tu + (1-t)v \in W(A)$ for all $t \in [0, 1]$.

For any complex numbers α and β, it is easy to verify that

$$W(\alpha I + \beta A) = \{\alpha + \beta z : z \in W(A)\}.$$

Intuitively the convexity of $W(A)$ does not change under shifting, scaling, and rotation. Thus, we may assume that the two points to be considered are 0 and 1, and show that $[0, 1] \subseteq W(A)$. Write

$$A = H + iK,$$

where
$$H = \frac{1}{2}(A + A^*) \text{ and } K = \frac{1}{2i}(A - A^*)$$
are Hermitian matrices. Let x and y be unit vectors in $\mathbb{C}^n$ such that
$$x^* A x = 0, \quad y^* A y = 1.$$
It follows that x and y are linearly independent and that
$$x^* H x = x^* K x = y^* K y = 0, \quad y^* H y = 1.$$
We may further assume that $x^* K y$ has real part zero; otherwise, one may replace x with cx, $c \in \mathbb{C}$ and $|c| = 1$, so that $cx^* K y$ is 0 or a pure complex number without changing the value of $x^* A x$.

Note that $tx + (1-t)y \neq 0$, $t \in [0, 1]$. Define for $t \in [0, 1]$
$$z(t) = \frac{1}{\|tx + (1-t)y\|^2}(tx + (1-t)y).$$
Then $z(t)$ is a unit vector. It is easy to compute that for all $t \in [0, 1]$
$$z(t)^* K z(t) = 0.$$
The convexity of $W(A)$ then follows, for
$$\{z(t)^* A z(t) : 0 \le t \le 1\} = [0, 1].$$

The compactness of $W(A)$, meaning the boundary is contained in $W(A)$, is seen by noting that $W(A)$ is the range of the continuous function $x \mapsto x^* A x$ on the compact set $\{x \in \mathbb{C}^n : \|x\| = 1\}$. (A continuous function maps a compact set to a compact set.) ∎

When considering the smallest disc centered at the origin that covers the numerical range, we associate to $W(A)$ a number
$$w(A) = \sup\{|z| : z \in W(A)\},$$
called the *numerical radius* of $A \in \mathbb{M}_n$, that is,
$$w(A) = \sup_{\|x\|=1} |x^* A x|.$$

It is immediate that for any $x \in \mathbb{C}^n$

$$|x^*Ax| \leq w(A)\|x\|^2. \tag{3.8}$$

We now make comparisons of the numerical radius $w(A)$ to the largest eigenvalue $\rho(A)$ in absolute value, or the *spectral radius*, i.e.,

$$\rho(A) = \max\{|\lambda| : \lambda \text{ is an eigenvalue of } A\},$$

and to the largest singular value $\sigma_{\max}(A)$, also called the *spectral norm*. It is easy to see (Problem 7) that

$$\sigma_{\max}(A) = \sup_{\|x\|=1} \|Ax\| = \sup_{x \neq 0} \frac{\|Ax\|}{\|x\|}$$

and that for every $x \in \mathbb{C}^n$

$$\|Ax\| \leq \sigma_{\max}(A)\|x\|.$$

Theorem 3.16 *Let A be a square complex matrix. Then*

$$\rho(A) \leq w(A) \leq \sigma_{\max}(A) \leq 2w(A).$$

PROOF. Let λ be the eigenvalue of A such that $\rho(A) = |\lambda|$, and let u be a unit eigenvector corresponding to λ. Then

$$\rho(A) = |u^*Au| \leq w(A).$$

The second inequality follows from the Cauchy-Schwarz inequality

$$|x^*Ax| = |(Ax, x)| \leq \|Ax\|\|x\|.$$

We have left to show that $\sigma_{\max}(A) \leq 2w(A)$. It can be verified that

$$\begin{aligned}4(Ax, y) &= \Big(A(x+y), x+y\Big) - \Big(A(x-y), x-y\Big) \\&\quad + i\Big(A(x+iy), x+iy\Big) - i\Big(A(x-iy), x-iy\Big).\end{aligned}$$

Using (3.8), it follows that

$$\begin{aligned}4|(Ax, y)| &\leq w(A)(\|x+y\|^2 + \|x-y\|^2 \\&\quad + \|x+iy\|^2 + \|x-iy\|^2) \\&= 4w(A)(\|x\|^2 + \|y\|^2).\end{aligned}$$

Thus, for any unit x and y in $\mathbb{C}^n$, we have

$$|(Ax, y)| \leq 2w(A).$$

The inequality follows immediately from Problem 8. ∎

Problems ─────────────────────────────────────

1. Find a nonzero matrix A so that $\rho(A) = 0$.

2. Find the eigenvalues, singular values, numerical radius, spectral radius, spectral norm, and numerical range for each of the following:

$$\begin{pmatrix} 1 & 1 \\ 0 & 0 \end{pmatrix}, \begin{pmatrix} 0 & 1 \\ 1 & 0 \end{pmatrix}, \begin{pmatrix} 1 & 1 \\ 0 & 1 \end{pmatrix}, \begin{pmatrix} 1 & 1 \\ 1 & 1 \end{pmatrix}.$$

3. Let A be an n-square complex matrix. Show that the numerical radius, spectral radius, spectral norm, and numerical range are unitarily invariant. That is, for instance, $w(U^*AU) = w(A)$ for any n-square unitary matrix U.

4. Show that the diagonal entries and the eigenvalues of a square matrix are contained in the numerical range of the matrix.

5. Let $A \in \mathbb{M}_n$. Show that $\frac{1}{n}\operatorname{tr} A$ is contained in $W(A)$. Conclude that for any nonsingular $P \in \mathbb{M}_n$, $W(P^{-1}AP - PAP^{-1})$ contains 0.

6. Let A be a square complex matrix. Show that $\frac{\|Ax\|}{\|x\|}$ is constant for all $x \neq 0$ if and only if all the singular values of A are identical.

7. Let A be a complex matrix. Show that

$$\sigma_{\max}(A) = \sqrt{\rho(A^*A)} = \sup_{\|x\|=1} \|Ax\| = \sup_{x \neq 0} \frac{\|Ax\|}{\|x\|}.$$

8. Let A be an n-square complex matrix. Show that

$$\sigma_{\max}(A) = \sup_{\|x\|=\|y\|=1} |(Ax, y)|.$$

9. Show that for any matrices A and B of the same size,

$$\sigma_{\max}(A + B) \leq \sigma_{\max}(A) + \sigma_{\max}(B).$$

10. Show that the numerical range of a 2×2 matrix is in general a closed elliptical disc (possibly degenerate as a singleton or a line segment).

11. Take
$$A = \begin{pmatrix} 0 & 1 & 0 & 0 \\ 0 & 0 & 1 & 0 \\ 0 & 0 & 0 & 1 \\ 0 & 0 & 0 & 0 \end{pmatrix}$$
and let $B = A^2$. Find $w(A)$, $w(B)$, and $w(AB)$.

12. Show that the numerical range of a normal matrix is the *convex hull* of its eigenvalues. That is, if $A \in \mathbb{M}_n$ is a normal matrix with eigenvalues $\lambda_1, \ldots, \lambda_n$, then
$$W(A) = \{t_1\lambda_1 + \cdots + t_n\lambda_n : t_1 + \cdots + t_n = 1, \text{ each } t_i \geq 0\}.$$

13. Show that $W(A)$ is a polygon inscribed in the unit circle if A is unitary, and that $W(A) \subseteq \mathbb{R}$ if A is Hermitian. What can be said about $W(A)$ if A is positive semidefinite?

14. Show that $w(A) = \rho(A) = \sigma_{\max}(A)$ if A is normal. Discuss the converse by considering
$$A = \text{diag}(1, i, -1, -i) \oplus \begin{pmatrix} 0 & 1 \\ 0 & 0 \end{pmatrix}.$$

15. Prove or disprove that for any n-square complex matrices A and B
    (a) $\rho(AB) \leq \rho(A)\rho(B)$,
    (b) $w(AB) \leq w(A)w(B)$,
    (c) $\sigma_{\max}(AB) \leq \sigma_{\max}(A)\sigma_{\max}(B)$.

16. Let A be a square matrix. Show that for every positive integer k
$$w(A^k) \leq \left(w(A)\right)^k.$$

Is it true in general that
$$w(A^{k+m}) \leq w(A^k)\, w(A^m)?$$

CHAPTER 4

Special Types of Matrices

INTRODUCTION This chapter studies special types of matrices. They are: idempotent matrices, nilpotent matrices, involutary matrices, projection matrices, tridiagonal matrices, circulant matrices, Vandermonde matrices, Hadamard matrices, permutation matrices, and doubly stochastic matrices. These matrices are often used in many subjects of mathematics and in other fields.

4.1 Idempotence, Nilpotence, Involution, and Projections

We first present three types of matrices that have simple structures under similarity: idempotent matrices, nilpotent matrices, and involutions. We then turn attention to orthogonal projection matrices.

A square matrix A is said to be *idempotent*, or a *projection*, if
$$A^2 = A,$$
nilpotent if for some positive integer k
$$A^k = 0,$$
and *involutary* if
$$A^2 = I.$$

Theorem 4.1 *Let A be an n-square complex matrix. Then*

1. *A is idempotent if and only if A is similar to a diagonal matrix of the form* $\operatorname{diag}(1, \ldots, 1, 0, \ldots, 0)$,

2. *A is nilpotent if and only if all the eigenvalues of A are zero,*

3. *A is involutary if and only if A is similar to a diagonal matrix of the form* $\operatorname{diag}(1, \ldots, 1, -1, \ldots, -1)$.

PROOF. The sufficiency in (1) is obvious. To see the necessity, let

$$A = P^{-1}(J_1 \oplus \cdots \oplus J_k)P$$

be a Jordan decomposition of A. Then for each i, $i = 1, \ldots, k$

$$A^2 = A \quad \Rightarrow \quad J_i^2 = J_i.$$

Observe that if J is a Jordan block and if $J^2 = J$, then J must be of size 1, that is, J is a number. The assertion then follows.

For (2), consider the Schur (or Jordan) decomposition of A,

$$A = U^{-1} \begin{pmatrix} \lambda_1 & & * \\ & \ddots & \\ 0 & & \lambda_n \end{pmatrix} U,$$

where U is an n-square unitary matrix.

If $A^k = 0$, then each $\lambda_i^k = 0$, and A has only zero eigenvalues. Conversely, it is easy to verify by computation that $A^n = 0$ if all the eigenvalues of A are equal to zero (see also Problem 10, Section 3.3).

The proof of (3) is similar to that of (1). ∎

Theorem 4.2 *Let A and B be nilpotent matrices of the same size. If A and B commute, then $A + B$ is nilpotent.*

PROOF. Let $A^m = 0$ and $B^n = 0$. Upon computation, we have

$$(A + B)^{m+n} = 0,$$

for each term in the expansion of $(A + B)^{m+n}$ is A^{m+n}, is B^{m+n}, or contains $A^s B^t$, $s \geq m$ or $t \geq n$. In any case, every term vanishes. ∎

SEC. 4.1   IDEMPOTENCE, NILPOTENCE, INVOLUTION, AND PROJECTIONS   95

By choosing a suitable basis for $\mathbb{C}^n$, we can interpret Theorem 4.1(1) as follows: A matrix A is a projection if and only if $\mathbb{C}^n$ can be decomposed as
$$\mathbb{C}^n = W_1 \oplus W_2, \qquad (4.1)$$
where W_1 and W_2 are subspaces such that for all $w_1 \in W_1$, $w_2 \in W_2$
$$Aw_1 = w_1, \quad Aw_2 = 0.$$
Thus, if $w = w_1 + w_2 \in \mathbb{C}^n$, where $w_1 \in W_1$ and $w_2 \in W_2$, then
$$Aw = Aw_1 + Aw_2 = w_1.$$
Such a w_1 is called the *projection* of w on W_1. Note that
$$W_1 = \operatorname{Im} A, \quad W_2 = \operatorname{Ker} A = \operatorname{Im}(I - A).$$
Using this and Theorem 4.1(1), one may prove the next result.

Theorem 4.3 *For any $A \in \mathbb{M}_n$ the following are equivalent:*

1. *A is a projection matrix, that is, $A^2 = A$;*
2. *$\mathbb{C}^n = \operatorname{Im} A + \operatorname{Ker} A$ with $Ax = x$ for every $x \in \operatorname{Im} A$;*
3. *$\operatorname{Ker} A = \operatorname{Im}(I - A)$;*
4. *$\operatorname{rank}(A) + \operatorname{rank}(I - A) = n$;*
5. *$\operatorname{Im} A \cap \operatorname{Im}(I - A) = \{0\}$.*

We now turn our attention to orthogonal projection matrices. A square complex matrix A is called an *orthogonal projection* if
$$A^2 = A = A^*.$$
For orthogonal projection matrices, the subspaces
$$W_1 = \operatorname{Im} A \quad \text{and} \quad W_2 = \operatorname{Im}(I - A)$$
in (4.1) are orthogonal, that is, for all $w_1 \in W_1$ and $w_2 \in W_2$,
$$(w_1, w_2) = 0. \qquad (4.2)$$
In other words, $(Ax, (I - A)x) = 0$ for all $x \in \mathbb{C}^n$; this is because
$$(w_1, w_2) = (Aw_1, w_2) = (w_1, A^* w_2) = (w_1, Aw_2) = 0.$$

Theorem 4.4 *For any $A \in \mathbb{M}_n$ the following are equivalent:*

1. *A is an orthogonal projection matrix, that is, $A^2 = A = A^*$;*

2. *$A = U^* \operatorname{diag}(1, \cdots, 1, 0, \cdots, 0) U$ for some unitary matrix U;*

3. *$\|x - Ax\| \leq \|x - Ay\|$ for every x and y in $\mathbb{C}^n$;*

4. *$A^2 = A$ and $\|Ax\| \leq \|x\|$ for every $x \in \mathbb{C}^n$;*

5. *$A = A^* A$.*

PROOF. (1)⇔(2): We show (1)⇒(2). The other direction is obvious.

Since A is Hermitian, by the spectral decomposition theorem (Theorem 3.4), we have $A = V^* \operatorname{diag}(\lambda_1, \ldots, \lambda_n) V$ for some unitary matrix V, where the λ_i are the eigenvalues of A. However, A is idempotent and thus has only eigenvalues 1 and 0 according to the previous theorem. It follows that

$$A = U^* \operatorname{diag}(\overbrace{1, \ldots, 1}^{r}, 0, \ldots, 0) U,$$

where r is the rank of A and U is some unitary matrix.

(1)⇔(3): For (1)⇒(3), let A be an orthogonal projection. We have the decomposition (4.1) with the orthogonality condition (4.2). Let $x = x_1 + x_2$, where $x_1 \in W_1$, $x_2 \in W_2$, and $(x_1, x_2) = 0$. Similarly, write $y = y_1 + y_2$. Note that $x_1 - y_1 \in W_1$ and $W_1 \perp W_2$. Using the fact that $(u, v) = 0 \Rightarrow \|u\| + \|v\| = \|u + v\|$, we have

$$\|x - Ax\| = \|x_2\| \leq \|x_2\| + \|x_1 - y_1\| = \|x_2 + (x_1 - y_1)\| = \|x - Ay\|.$$

We now show (3)⇒(1). It is sufficient to show that the decomposition (4.1) with the orthogonality condition (4.2) holds, where Im A serves as W_1 and $\operatorname{Im}(I - A)$ as W_2.

Since $x = Ax + (I - A)x$ for every $x \in \mathbb{C}^n$, it is obvious that

$$\mathbb{C}^n = \operatorname{Im} A + \operatorname{Im}(I - A).$$

We have left to show that $(x, y) = 0$ if $x \in \operatorname{Im} A$ and $y \in \operatorname{Im}(I - A)$. Suppose instead that $((I - A)x, Ay) \neq 0$ for some x and $y \in \mathbb{C}^n$. We shall show that there exists a vector $z \in \mathbb{C}^n$ such that

$$\|x - Az\| < \|x - Ax\|,$$

SEC. 4.1   IDEMPOTENCE, NILPOTENCE, INVOLUTION, AND PROJECTIONS   97

which is a contradiction to the given condition (3).

Let $((I - A)x, Ay) = \alpha \neq 0$. We may assume that $\alpha < 0$. Otherwise, replace x with $e^{i\theta}x$, where $\theta \in \mathbb{R}$ is such that $e^{i\theta}\alpha < 0$.

Let $z_\epsilon = x - \epsilon y$, where $\epsilon > 0$. Then

$$\begin{aligned}\|x - Az_\epsilon\|^2 &= \|(x - Ax) + (Ax - Az_\epsilon)\|^2 \\ &= \|x - Ax\|^2 + \|Ax - Az_\epsilon\|^2 \\ &\quad + 2\operatorname{Re}((I - A)x, A(x - z_\epsilon)) \\ &= \|x - Ax\|^2 + \|Ax - Az_\epsilon\|^2 \\ &\quad + 2\epsilon((I - A)x, Ay) \\ &= \|x - Ax\|^2 + \epsilon^2\|Ay\|^2 + 2\epsilon\alpha.\end{aligned}$$

Since $\alpha < 0$, we have $\epsilon^2\|Ay\|^2 + 2\epsilon\alpha < 0$ for some ϵ small enough, which results in a contradiction to the assumption in (3):

$$\|x - Az_\epsilon\| < \|x - Ax\|.$$

(1)$\Rightarrow$(4): If A is an orthogonal projection matrix, then the orthogonality condition (4.2) holds. Thus, $(Ax, (I - A)x) = 0$ and

$$\|Ax\|^2 \leq \|Ax\|^2 + \|(I - A)x\|^2 = \|Ax + (I - A)x\|^2 = \|x\|^2.$$

(4)$\Rightarrow$(5): If $A \neq A^*A$, that is, $(A^* - I)A \neq 0$ or $A^*(I - A) \neq 0$, then $\operatorname{rank}(I - A) < n$ and $\dim \operatorname{Im}(I - A) < n$ by Theorem 4.1(1).

We show that there exists a nonzero x such that

$$(x, (I - A)x) = 0, \quad \text{but} \quad (I - A)x \neq 0.$$

Thus, for this x

$$\|Ax\|^2 = \|x - (I - A)x\|^2 = \|x\|^2 + \|(I - A)x\|^2 > \|x\|^2,$$

which contradicts the condition $\|Ax\| \leq \|x\|$ for every $x \in \mathbb{C}^n$.

To show the existence of such a vector x, it is sufficient to show that there exists a nonzero x in $(\operatorname{Im}(I - A))^\perp$ but not in $\operatorname{Ker}(I - A)$, that is, $(\operatorname{Im}(I - A))^\perp$ is not contained in $\operatorname{Ker}(I - A)$.

Notice that (Theorem 1.4)

$$\dim \operatorname{Im}(I - A) + \dim \operatorname{Ker}(I - A) = n$$

and that
$$\mathbb{C}^n = \operatorname{Im}(I - A) \oplus (\operatorname{Im}(I - A))^\perp.$$

Now if $(\operatorname{Im}(I - A))^\perp$ is contained in $\operatorname{Ker}(I - A)$, then they must be equal, for they have the same dimension:

$$\dim(\operatorname{Im}(I - A))^\perp = n - \dim \operatorname{Im}(I - A) = \dim \operatorname{Ker}(I - A).$$

It follows, by (1.11) in Section 1.4 of Chapter 1, that

$$\operatorname{Im}(I - A) = \operatorname{Im}(I - A^*).$$

Thus, $I - A = I - A^*$ and A is Hermitian. Then (5) follows easily. (5)$\Rightarrow$(1): If $A = A^*A$, then A is obviously Hermitian. Thus,

$$A = A^*A = AA = A^2. \quad \blacksquare$$

Problems

1. Characterize all 2×2 idempotent, nilpotent, and involutary matrices up to similarity.

2. Can a nonzero matrix be both idempotent and nilpotent? Why?

3. What is the characteristic polynomial of a nilpotent matrix?

4. What idempotent matrices are nonsingular?

5. Show that if $A \in \mathbb{M}_n$ is idempotent, then so is $P^{-1}AP$ for any invertible $P \in \mathbb{M}_n$.

6. Show that the rank of an idempotent matrix is equal to the number of nonzero eigenvalues of the matrix.

7. Let A and B be idempotent matrices of the same size. Find the necessary and sufficient conditions for $A + B$ to be idempotent. Discuss the analog for $A - B$.

8. Show that $\frac{1}{2}(I + A)$ is idempotent if and only if A is an involution.

9. Let A be a square complex matrix. Show that

$$A^2 = A \quad \Leftrightarrow \quad \operatorname{rank}(A) = \operatorname{tr}(A) \text{ and } \operatorname{rank}(I - A) = \operatorname{tr}(I - A).$$

SEC. 4.1   IDEMPOTENCE, NILPOTENCE, INVOLUTION, AND PROJECTIONS   99

10. Let A be an idempotent matrix. Show that
$$A = A^* \quad \Leftrightarrow \quad \operatorname{Im} A = \operatorname{Im} A^*.$$

11. Show that a Hermitian idempotent matrix is positive semidefinite and that the matrix M is positive semidefinite, where
$$M = I - \frac{1}{y^*y} yy^*, \quad y \in \mathbb{C}^n.$$

12. Show that $\mathcal{T} = \mathcal{T}^2$, where $\mathcal{T}$ is a transformation defined on $\mathbb{M}_2$ by
$$\mathcal{T}(X) = TX - XT, \quad X \in \mathbb{M}_2,$$
with
$$T = \begin{pmatrix} 0 & 1 \\ 0 & 0 \end{pmatrix}.$$
Find the Jordan form of a matrix representation of $\mathcal{T}$.

13. Let A and B be square matrices of the same size. If A is nilpotent and $AB = BA$, show that AB is nilpotent. Is the converse true?

14. Let A be an n-square nonsingular matrix. If X is a matrix such that
$$AXA^{-1} = \lambda X, \quad \lambda \in \mathbb{C},$$
show that $|\lambda| = 1$ or X is nilpotent.

15. Give a 2×2 matrix such that $A^2 = I$ but $A^*A \neq I$.

16. Let $A^2 = A$. Show that $(A+I)^k = I + (2^k - 1)A$ for $k = 1, 2, \ldots$.

17. Find a matrix that is a projection but not an orthogonal projection.

18. Let A and B be square matrices of the same size. If $AB = A$ and $BA = B$, show that A and B are projection matrices.

19. Let A be a projection matrix. Show that A is Hermitian if and only if $\operatorname{Im} A$ and $\operatorname{Ker} A$ are orthogonal, that is, $\operatorname{Im} A \perp \operatorname{Ker} A$.

20. Prove Theorem 4.3 along the line: $(1) \Leftrightarrow (2) \Rightarrow (3) \Rightarrow (4) \Rightarrow (5) \Rightarrow (2)$.

21. Show that A is an orthogonal projection matrix if and only if $A = B^*B$ for some matrix B with $BB^* = I$.

22. Let $A_1, \ldots, A_m$ be $n \times n$ idempotent matrices. If
$$A_1 + \cdots + A_m = I_n,$$
show that
$$A_i A_j = 0, \quad i \neq j.$$
(Hint: Show that $\mathbb{C}^n = \operatorname{Im} A_1 \oplus \cdots \oplus \operatorname{Im} A_m$ by using trace.)

23. If W is a subspace of an inner product space V, one may write
$$V = W \oplus W^\perp$$
and define a transformation $\mathcal{A}$ on V by
$$\mathcal{A}(v) = w, \quad \text{if } v = w + w^\perp, \ w \in W, \ w^\perp \in W^\perp,$$
where w is called the *projection* of v on W. Show that

(a) $\mathcal{A}$ is a linear transformation,
(b) $\mathcal{A}^2 = \mathcal{A}$,
(c) $\text{Im}(\mathcal{A}) = W$ and $\text{Ker}(\mathcal{A}) = W^\perp$,
(d) $\|v - \mathcal{A}(v)\| \leq \|v - \mathcal{A}(u)\|$ for any $u \in V$,
(e) Every $v \in V$ has a unique projection $w \in W$,
(f) $\|v\| = \|w\| + \|w^\perp\|$.

24. When does equality in Theorem 4.4(3) hold?

25. Let A and B be $m \times n$ complex matrices of rank n, $n \leq m$. Show that the matrix $A(A^*A)^{-1}A^*$ is idempotent and that
$$A(A^*A)^{-1}A^* = B(B^*B)^{-1}B^* \iff A = BX$$
for some nonsingular matrix X. (Hint: Multiply by A.)

26. Let A and B be orthogonal projections of the same size. Show that $A + B$ is an orthogonal projection if and only if $AB = BA = 0$.

27. Let A and B be orthogonal projections of the same size. Show that $A - B$ is an orthogonal projection if and only if $AB = BA = B$.

28. Let A and B be orthogonal projections of the same size. Show that AB is an orthogonal projection if and only if $AB = BA$.

4.2 Tridiagonal Matrices

One of the frequently used techniques in determinantal computation is recursion. We illustrate this method by computing the determinant of a tridiagonal matrix and go on studying the eigenvalues of matrices of this kind.

An n-square *tridiagonal matrix*, symbolized as T_n, is a matrix with entries $t_{ij} = 0$ when $|i - j| > 1$. For simplicity, we consider the special tridiagonal matrix

$$T_n = \begin{pmatrix} a & b & & & & & 0 \\ c & a & b & & & & \\ & c & a & b & & & \\ & & \ddots & \ddots & \ddots & & \\ & & & & c & a & b \\ 0 & & & & & c & a \end{pmatrix}. \qquad (4.3)$$

Theorem 4.5 *Let T_n be defined as in (4.3). Then*

$$\det T_n = \begin{cases} a^n & \text{if } bc = 0, \\ (n+1)(a/2)^n & \text{if } a^2 = 4bc, \\ (\alpha^{n+1} - \beta^{n+1})/(\alpha - \beta) & \text{if } a^2 \neq 4bc, \end{cases}$$

where

$$\alpha = \frac{a + \sqrt{a^2 - 4bc}}{2}, \quad \beta = \frac{a - \sqrt{a^2 - 4bc}}{2}.$$

PROOF. Expand the determinant along the first row of the matrix in (4.3) to obtain the recursive formula

$$\det T_n = a \det T_{n-1} - bc \det T_{n-2}. \qquad (4.4)$$

If $bc = 0$, then $b = 0$ or $c = 0$, and from (4.3) obviously $\det T_n = a^n$. If $bc \neq 0$, let α and β be the solutions to $x^2 - ax + bc = 0$. Then

$$\alpha + \beta = a, \quad \alpha\beta = bc.$$

Note that
$$a^2 - 4bc = (\alpha - \beta)^2.$$
From the recursive formula (4.4), we have
$$\det T_n - \alpha \det T_{n-1} = \beta(\det T_{n-1} - \alpha \det T_{n-2})$$
and
$$\det T_n - \beta \det T_{n-1} = \alpha(\det T_{n-1} - \beta \det T_{n-2}).$$
Denote
$$f_n = \det T_n - \alpha \det T_{n-1}, \quad g_n = \det T_n - \beta \det T_{n-1}.$$
Then
$$f_n = \beta f_{n-1}, \quad g_n = \alpha g_{n-1},$$
with, by a simple computation,
$$f_2 = \beta^2, \quad g_2 = \alpha^2.$$
Thus,
$$f_n = \beta^n, \quad g_n = \alpha^n,$$
that is,
$$\det T_n - \alpha \det T_{n-1} = \beta^n, \quad \det T_n - \beta \det T_{n-1} = \alpha^n. \qquad (4.5)$$
It follows, using T_{n+1} in (4.5) and subtracting the equations, that
$$\det T_n = \frac{\alpha^{n+1} - \beta^{n+1}}{\alpha - \beta}, \quad \text{if } \alpha \neq \beta,$$
and, by the induction hypothesis, that
$$\det T_n = (n+1)\left(\frac{a}{2}\right)^n, \quad \text{if } \alpha = \beta. \blacksquare$$

Note that the recursive formula (4.4) in the proof depends not on the single values of b and c but on the product bc. Thus, if $a \in \mathbb{R}$, $bc > 0$, we may replace b and c by d and $\bar{d}$, respectively, where $d\bar{d} = bc$, to get a tridiagonal Hermitian matrix H_n, for which
$$\det(\lambda I - T_n) = \det(\lambda I - H_n).$$
It follows that T_n has only real eigenvalues because H_n does. In fact, when $a, b, c \in \mathbb{R}$ and $bc > 0$, matrix DT_nD^{-1} is real symmetric, where D is the diagonal matrix $\text{diag}(1, e, \ldots, e^{n-1})$ with $e = \sqrt{b/c}$.

Theorem 4.6 *If T_n is a tridiagonal matrix defined as in (4.3) with $a \in \mathbb{R}$ and $bc > 0$, then the eigenvalues of T_n are all real and have eigenspaces of dimension one.*

PROOF. The first half follows from the argument prior to the theorem. For the second part, it is sufficient to prove that each eigenvalue has only one eigenvector up to a factor.

Let $x = (x_1, \ldots, x_n)^T$ be an eigenvector of T_n corresponding to the eigenvalue λ. Then
$$(\lambda I - T_n)x = 0, \quad x \neq 0,$$
or equivalently
$$\begin{aligned}
(\lambda - a)x_1 - bx_2 &= 0, \\
-cx_1 + (\lambda - a)x_2 - bx_3 &= 0, \\
&\vdots \\
-cx_{n-2} + (\lambda - a)x_{n-1} - bx_n &= 0, \\
-cx_{n-1} + (\lambda - a)x_n &= 0.
\end{aligned}$$

Since $b \neq 0$, x_2 is determined by x_1 in the first equation, so are $x_3, \ldots, x_n$ successively by x_2, x_3, and so on in the equations 2, 3, $\ldots$, $n-1$. If x_1 is replaced by kx_1, then $x_2, x_3, \ldots, x_n$ become $kx_2, kx_3, \ldots, kx_n$, and the eigenvector is unique up to a factor. ∎

Note that the theorem is in fact true for a general tridiagonal matrix when a_i is real and $b_i c_i > 0$ for each i.

Problems

1. Compute the determinant
$$\begin{vmatrix} a & b & 0 \\ c & a & b \\ 0 & c & a \end{vmatrix}.$$

2. Carry out in detail the proof that T_n is similar to a real symmetric matrix if $a, b, c \in \mathbb{R}$ and $bc > 0$.

3. Compute the $n \times n$ determinant

$$\begin{vmatrix} 0 & 1 & & & & & 0 \\ 1 & 0 & 1 & & & & \\ & 1 & 0 & 1 & & & \\ & & \ddots & \ddots & \ddots & & \\ & & & 1 & 0 & 1 \\ 0 & & & & & 1 & 0 \end{vmatrix}.$$

4. Compute the $n \times n$ determinant

$$\begin{vmatrix} 1 & 1 & & & & & 0 \\ -1 & 1 & 1 & & & & \\ & -1 & 1 & 1 & & & \\ & & \ddots & \ddots & \ddots & & \\ & & & -1 & 1 & 1 \\ 0 & & & & & -1 & 1 \end{vmatrix}.$$

5. Compute the $n \times n$ determinant

$$\begin{vmatrix} 2 & 1 & & & & & 0 \\ 1 & 2 & 1 & & & & \\ & 1 & 2 & 1 & & & \\ & & \ddots & \ddots & \ddots & & \\ & & & 1 & 2 & 1 \\ 0 & & & & & 1 & 2 \end{vmatrix}.$$

6. Compute the $n \times n$ determinant

$$\begin{vmatrix} 3 & 2 & & & & & 0 \\ 1 & 3 & 1 & & & & \\ & 2 & 3 & 2 & & & \\ & & 1 & 3 & 1 & & \\ & & & \ddots & \ddots & \ddots & \\ & & & & 1 & 3 & 1 \\ 0 & & & & & 2 & 3 \end{vmatrix}.$$

7. Find the inverse of the $n \times n$ matrix

$$\begin{pmatrix} 2 & -1 & & & & & 0 \\ -1 & 2 & -1 & & & & \\ & -1 & 2 & -1 & & & \\ & & -1 & 2 & -1 & & \\ & & & \ddots & \ddots & \ddots & \\ & & & & -1 & 2 & -1 \\ 0 & & & & & -1 & 2 \end{pmatrix}.$$

8. Show that the value of the following determinant is independent of x:

$$\begin{vmatrix} a & x & & & & & 0 \\ \frac{1}{x} & a & x & & & & \\ & \frac{1}{x} & a & x & & & \\ & & \ddots & \ddots & \ddots & & \\ & & & \frac{1}{x} & a & x & \\ 0 & & & & \frac{1}{x} & a \end{vmatrix}.$$

9. (**Cauchy matrix**) Let $\lambda_1, \ldots, \lambda_n$ be positive numbers and let

$$\Lambda = \left(\frac{1}{\lambda_i + \lambda_j} \right).$$

Show that

$$\det \Lambda = \frac{\prod_{i>j}(\lambda_i - \lambda_j)^2}{\prod_{i,j}(\lambda_i + \lambda_j)}.$$

(Hint: Subtract the last column from each of the other columns, then factor; do the same thing for rows; use induction.)

10. Show that

$$\begin{vmatrix} \frac{1}{2} & \frac{1}{3} & \cdots & \frac{1}{n+1} \\ \frac{1}{3} & \frac{1}{4} & \cdots & \frac{1}{n+2} \\ \vdots & \vdots & \vdots & \vdots \\ \frac{1}{n+1} & \frac{1}{n+2} & \cdots & \frac{1}{2n} \end{vmatrix} \geq 0.$$

4.3 Circulant Matrices

An n-square *circulant matrix* is a matrix of the form

$$\begin{pmatrix} c_0 & c_1 & c_2 & \cdots & c_{n-1} \\ c_{n-1} & c_0 & c_1 & \cdots & c_{n-2} \\ c_{n-2} & c_{n-1} & c_0 & \cdots & c_{n-3} \\ \vdots & \vdots & \vdots & \vdots & \vdots \\ c_1 & c_2 & c_3 & \cdots & c_0 \end{pmatrix}, \qquad (4.6)$$

where $c_0, c_1, \ldots, c_{n-1}$ are complex numbers. For instance,

$$N = \begin{pmatrix} 1 & 2 & 3 & \cdots & n \\ n & 1 & 2 & \cdots & n-1 \\ \vdots & \vdots & \vdots & \vdots & \vdots \\ 3 & 4 & 5 & \cdots & 2 \\ 2 & 3 & 4 & \cdots & 1 \end{pmatrix}$$

and

$$P = \begin{pmatrix} 0 & 1 & 0 & \cdots & 0 \\ 0 & 0 & 1 & \cdots & 0 \\ \vdots & \vdots & \vdots & \vdots & \vdots \\ 0 & 0 & 0 & \cdots & 1 \\ 1 & 0 & 0 & \cdots & 0 \end{pmatrix} \qquad (4.7)$$

are circulant matrices. Note that P is also a permutation matrix. We refer to this P as the $n \times n$ *primary permutation matrix*.

This section deals with the basic properties of circulant matrices. The following theorem may be shown by a direct verification.

Theorem 4.7 *An n-square matrix C is circulant if and only if*

$$C = PCP^T,$$

where P is the $n \times n$ primary permutation matrix.

SEC. 4.3   CIRCULANT MATRICES   107

We call a complex number ω an nth *primitive root* of unity if $\omega^n - 1 = 0$ and $\omega^k - 1 \neq 0$ for every positive integer $k < n$.

Note that if ω is an nth primitive root of unity, then ω^k is a solution to $x^n - 1 = 0$, $0 < k < n$. It follows by factoring $x^n - 1$ that

$$\sum_{i=0}^{n-1} \omega^{ik} = \frac{(\omega^k)^n - 1}{\omega^k - 1} = 0.$$

Theorem 4.8 *Let C be a circulant matrix in the form (4.6), and let $f(\lambda) = c_0 + c_1 \lambda + \cdots + c_{n-1} \lambda^{n-1}$. Then*

1. *$C = f(P)$, where P is the $n \times n$ primary permutation matrix,*

2. *C is a normal matrix, that is, $C^* C = C C^*$,*

3. *the eigenvalues of C are $f(\omega^k)$, $k = 0, 1, \ldots, n-1$,*

4. *$\det C = f(\omega^0) f(\omega^1) \cdots f(\omega^{n-1})$,*

5. *$F^* C F$ is a diagonal matrix, where F is the unitary matrix with the (i, j)-entry equal to $\frac{1}{\sqrt{n}} \omega^{(i-1)(j-1)}$, $i, j = 1, \ldots, n$.*

PROOF. (1) is easy to see by a direct computation. (2) is due to the fact that if matrices A and B commute, so do $p(A)$ and $q(B)$, where p and q are any polynomials (Problem 4). Note that $PP^* = P^* P$.

For (3) and (4), the characteristic polynomial of P is

$$\det(\lambda I - P) = \lambda^n - 1 = \prod_{k=0}^{n-1} (\lambda - \omega^k).$$

Thus, the eigenvalues of P and P^i are, respectively, ω^k and ω^{ik}, $k = 0, 1, \ldots, n-1$. It follows that the eigenvalues of $C = f(P)$ are $f(\omega^k)$, $k = 0, 1, \ldots, n-1$ (Problem 6, Section 3.2), and that

$$\det C = \prod_{k=0}^{n-1} f(\omega^k).$$

To show (5), for each $k = 0, 1, \ldots, n-1$, let

$$x_k = (1, \omega^k, \omega^{2k}, \ldots, \omega^{(n-1)k})^T.$$

Then
$$Px_k = (\omega^k, \omega^{2k}, \ldots, \omega^{(n-1)k}, 1)^T = \omega^k x_k$$
and
$$Cx_k = f(P)x_k = f(\omega^k)x_k.$$

In other words, x_k are the eigenvectors of P and C corresponding to the eigenvalues ω^k and $f(\omega^k)$, respectively, $k = 0, 1, \ldots, n-1$.

However, since
$$(x_i, x_j) = \sum_{k=0}^{n-1} \overline{\omega^{jk}} \omega^{ik} = \sum_{k=0}^{n-1} \omega^{(i-j)k} = \begin{cases} 0, & i \neq j, \\ n, & i = j, \end{cases}$$

we have that
$$\left\{ \frac{1}{\sqrt{n}} x_0, \frac{1}{\sqrt{n}} x_1, \ldots, \frac{1}{\sqrt{n}} x_{n-1} \right\}$$

is an orthonormal basis for $\mathbb{C}^n$. Thus, we get a unitary matrix
$$F = \frac{1}{\sqrt{n}} \begin{pmatrix} 1 & 1 & 1 & \cdots & 1 \\ 1 & \omega & \omega^2 & \cdots & \omega^{n-1} \\ 1 & \omega^2 & \omega^4 & \cdots & \omega^{2(n-1)} \\ \vdots & \vdots & \vdots & \vdots & \vdots \\ 1 & \omega^{n-1} & \omega^{2(n-1)} & \cdots & \omega^{(n-1)(n-1)} \end{pmatrix}$$

such that
$$F^*CF = \text{diag}(f(\omega^0), f(\omega^1), \ldots, f(\omega^{n-1})).$$

That F is a unitary matrix is verified by a direct computation. ∎

Note that F, called a *Fourier matrix*, is independent of C.

Problems ─────────────────────────────

1. Let ω be an nth primitive root of unity. Show that
    (a) $\omega \bar{\omega} = 1$,
    (b) $\bar{\omega}^k = \omega^{-k} = \omega^{n-k}$,
    (c) $1 + \omega + \cdots + \omega^{n-1} = 0$.

2. Let ω be an nth primitive root of unity. Show that ω^k is also an nth primitive root of unity if and only if $(n, k) = 1$, that is, n and k have no common positive divisors other than 1.

3. Show that if A is a circulant matrix, then so are A^*, A^k, and A^{-1} if the inverse exists.

4. Let A and B be square matrices of the same size. If $AB = BA$, show that $p(A)q(B) = q(B)p(A)$ for any polynomials p and q.

5. Let A and B be circulant matrices of the same size. Show that A and B commute and that AB is a circulant matrix.

6. Let A be a circulant matrix. Show that for every positive integer k
$$\text{rank}(A^k) = \text{rank}(A).$$

7. Find the eigenvalues of the circulant matrices, with $\omega^3 = 1$,
$$\begin{pmatrix} 1 & 2 & 3 \\ 3 & 1 & 2 \\ 2 & 3 & 1 \end{pmatrix}, \quad \begin{pmatrix} 1 & 1 & 1 \\ 1 & \omega & \omega^2 \\ 1 & \omega^2 & \omega \end{pmatrix}.$$

8. Find the eigenvalues and the eigenvectors of the circulant matrix
$$\begin{pmatrix} 0 & 1 & 0 & 0 \\ 0 & 0 & 1 & 0 \\ 0 & 0 & 0 & 1 \\ 1 & 0 & 0 & 0 \end{pmatrix}.$$
Find the matrix F that diagonalizes the above matrix.

9. Let P be the $n \times n$ primary permutation matrix. Show that
$$P^n = I, \quad P^T = P^{-1} = P^{n-1}.$$

10. Find a matrix X such that
$$\begin{pmatrix} c_0 & c_1 & c_2 \\ c_1 & c_2 & c_0 \\ c_2 & c_0 & c_1 \end{pmatrix} = X \begin{pmatrix} c_0 & c_1 & c_2 \\ c_2 & c_0 & c_1 \\ c_1 & c_2 & c_0 \end{pmatrix}.$$

11. Find an invertible matrix Q such that
$$Q^* \begin{pmatrix} 0 & 0 & 0 & 1 \\ 1 & 0 & 0 & 0 \\ 0 & 1 & 0 & 0 \\ 0 & 0 & 1 & 0 \end{pmatrix} Q = \begin{pmatrix} 0 & 1 & 0 & 0 \\ 0 & 0 & 1 & 0 \\ 0 & 0 & 0 & 1 \\ 1 & 0 & 0 & 0 \end{pmatrix}.$$

12. Let F be the $n \times n$ Fourier matrix. Show that
    (a) F is symmetric, namely, $F^T = F$,
    (b) $(F^*)^2 = F^2$ is a permutation matrix,
    (c) $(F^*)^3 = F$ and $F^4 = I$,
    (d) the eigenvalues of F are ± 1 and $\pm i$ with appropriate multiplicity (which is the number of times the eigenvalue repeats),
    (e) $F^* = n^{-\frac{1}{2}} V(1, \omega, \omega^2, \ldots, \omega^{n-1})$, where V stands for the Vandermonde matrix (see the next section),
    (f) if $F = R + iS$, where R and S are real, then $R^2 + S^2 = I$, $RS = SR$, and R and S are symmetric.

13. Let e_i be the column vectors of n components with the ith component 1 and 0 elsewhere, $i = 1, 2, \ldots, n$, $c_1, c_2, \ldots, c_n \in \mathbb{C}$, and let
    $$A = \text{diag}(c_1, c_2, \ldots, c_n) P$$
    where P is the $n \times n$ primary permutation matrix. Show that
    (a) $A e_i = c_{i-1} e_{i-1}$ for each $i = 1, 2, \ldots, n$, where $c_0 = c_n$, $e_0 = e_n$,
    (b) $A^n = cI$, where $c = c_1 c_2 \cdots c_n$,
    (c) $\det(I + A + \cdots + A^{n-1}) = (1 - c)^{n-1}$.

14. A matrix is called a *Toeplitz matrix* if all entries of the matrix are constant down the diagonals parallel to the diagonal. In symbols,
    $$A = \begin{pmatrix} a_0 & a_1 & a_2 & \cdots & a_n \\ a_{-1} & a_0 & a_1 & \cdots & a_{n-1} \\ a_{-2} & a_{-1} & a_0 & \ddots & \vdots \\ \vdots & \vdots & \ddots & \ddots & a_1 \\ a_{-n} & a_{-n+1} & \cdots & a_{-1} & a_0 \end{pmatrix}.$$
    For example, matrix $B = (b_{ij})$ with $b_{i,i+1} = 1$, $i = 1, 2, \ldots, n-1$, and 0 elsewhere, is a Toeplitz matrix. Show that a matrix A is a Toeplitz matrix if and only if A can be written in the form
    $$A = \sum_{k=1}^{n} a_{-k}(B^T)^k + \sum_{k=0}^{n} a_k B^k.$$
    Show also that the sum of two Toeplitz matrices is a Toeplitz matrix and that a circulant matrix is Toeplitz.

4.4 Vandermonde Matrices

An n-square *Vandermonde matrix* is a matrix of the form

$$\begin{pmatrix} 1 & 1 & 1 & \cdots & 1 \\ a_1 & a_2 & a_3 & \cdots & a_n \\ a_1^2 & a_2^2 & a_3^2 & \cdots & a_n^2 \\ \vdots & \vdots & \vdots & \vdots & \vdots \\ a_1^{n-1} & a_2^{n-1} & a_3^{n-1} & \cdots & a_n^{n-1} \end{pmatrix},$$

denoted by $V_n(a_1, a_2, \ldots, a_n)$ or simply V.

Vandermonde matrices play a role in many places such as interpolation problems and solving systems of linear equations. We consider the determinant and the inverse of a Vandermonde matrix in this section.

Theorem 4.9 *Let $V_n(a_1, a_2, \ldots, a_n)$ be a Vandermonde matrix. Then*

$$\det V_n(a_1, a_2, \ldots, a_n) = \prod_{1 \leq i < j \leq n} (a_j - a_i),$$

and $V_n(a_1, a_2, \ldots, a_n)$ is invertible if and only if all the a_i are distinct.

PROOF. We proceed the proof by induction. There is nothing to show if $n = 1$ or 2. Let $n \geq 3$.

Suppose the assertion is true when the size of the matrix is $n - 1$. For the case of n, subtracting row i multiplied by a_1 from row $i + 1$, for i going down from $n - 1$ to 1, we have

$$\det V = \begin{vmatrix} 1 & 1 & 1 & \cdots & 1 \\ 0 & a_2 - a_1 & a_3 - a_1 & \cdots & a_n - a_1 \\ 0 & a_2(a_2 - a_1) & a_3(a_3 - a_1) & \cdots & a_n(a_n - a_1) \\ \vdots & \vdots & \vdots & \vdots & \vdots \\ 0 & a_2^{n-2}(a_2 - a_1) & a_3^{n-2}(a_3 - a_1) & \cdots & a_n^{n-2}(a_n - a_1) \end{vmatrix}$$

$$= \begin{vmatrix} a_2 - a_1 & a_3 - a_1 & \cdots & a_n - a_1 \\ a_2(a_2 - a_1) & a_3(a_3 - a_1) & \cdots & a_n(a_n - a_1) \\ \vdots & \vdots & \vdots & \vdots \\ a_2^{n-2}(a_2 - a_1) & a_3^{n-2}(a_3 - a_1) & \cdots & a_n^{n-2}(a_n - a_1) \end{vmatrix}$$

$$= \prod_{j=2}^{n}(a_j - a_1) \det V_{n-1}(a_2, a_3, \ldots, a_n)$$

$$= \prod_{j=2}^{n}(a_j - a_1) \prod_{2 \le i < j \le n}(a_j - a_i) \text{ (by the hypothesis)}$$

$$= \prod_{1 \le i < j \le n}(a_j - a_i).$$

It is readily seen that the Vandermonde matrix is singular if and only if at least two of the a_i are equal. ∎

An interesting application follows: Let $A \in \mathbb{M}_n$. Then

$$A^n = 0 \quad \Leftrightarrow \quad \operatorname{tr} A^k = 0, \; k = 1, 2, \ldots, n.$$

Since $A^n = 0$, then A is nilpotent, thus, A has only zero eigenvalues, so does A^k for each k. For the other way around, let the eigenvalues of A be $\lambda_1, \lambda_2, \ldots, \lambda_n$. Then the trace identities imply

$$\lambda_1 + \lambda_2 + \cdots + \lambda_n = 0,$$
$$\lambda_1^2 + \lambda_2^2 + \cdots + \lambda_n^2 = 0,$$
$$\vdots$$
$$\lambda_1^n + \lambda_2^n + \cdots + \lambda_n^n = 0,$$

rewritten as

$$V_n(\lambda_1, \lambda_2, \ldots, \lambda_n)(\lambda_1, \lambda_2, \ldots, \lambda_n)^T = 0.$$

If all of the λ_i are distinct, then by the preceding theorem the Vandermonde matrix is nonsingular and the system of equations in $\lambda_1, \lambda_2, \ldots, \lambda_n$ has only the trivial solution $\lambda_1 = \lambda_2 = \cdots = \lambda_n = 0$. If some of the λ_i are identical, say $\lambda_1 = \lambda_2$, and none of the other two are the same, we then write the system as

$$V_{n-1}(\lambda_2, \ldots, \lambda_n)(2\lambda_2, \ldots, \lambda_n)^T = 0.$$

A similar argument will result in $\lambda_2 = \cdots = \lambda_n = 0$.

This idea applies to the *interpolation problem* of finding a polynomial $f(x)$ of degree at most $n-1$ satisfying

$$f(x_i) = y_i, \quad i = 1, 2, \ldots, n,$$

where the x_i and y_i are given constants (Problem 4).

Theorem 4.10 *For any integers $k_1 < k_2 < \cdots < k_n$, the quotient*

$$\frac{\det V_n(k_1, k_2, \ldots, k_n)}{\det V_n(1, 2, \ldots, n)}$$

is an integer.

PROOF. Let f_i be any monic polynomial of degree i for each i. The additive property of determinants (see Section 1.2) shows that

$$\begin{vmatrix} 1 & 1 & 1 & \cdots & 1 \\ f_1(k_1) & f_1(k_2) & f_1(k_3) & \cdots & f_1(k_n) \\ f_2(k_1) & f_2(k_2) & f_2(k_3) & \cdots & f_2(k_n) \\ \vdots & \vdots & \vdots & \vdots & \vdots \\ f_{n-1}(k_1) & f_{n-1}(k_2) & f_{n-1}(k_3) & \cdots & f_{n-1}(k_n) \end{vmatrix} \quad (4.8)$$

is the same as $\det V_n(k_1, k_2, \ldots, k_n)$. By taking, for any integer a,

$$f_i(a) = a(a-1)(a-2)\cdots(a-i+1) = i!\binom{a}{i},$$

we see that $f_i(a)$ is divisible by $(i-1)!$.

Factoring out $(i-1)!$ from row i, $i = 2, 3, \ldots, n$, we see that the determinant in (4.8), thus $\det V_n(k_1, k_2, \ldots, k_n)$, is divisible by the product $\prod_{i=1}^{n}(i-1)!$.

The proof is complete, for $\prod_{i=1}^{n}(i-1)! = \det V_n(1, 2, \ldots, n)$. ∎

We now turn our attention to the inverse of a Vandermonde matrix. Consider the polynomial in x given by the product

$$p(x) = (x + a_1)(x + a_2) \cdots (x + a_n),$$

where $a_1, a_2, \ldots, a_n$ are constants. Expand $p(x)$ as a polynomial

$$p(x) = s_0 x^n + s_1 x^{n-1} + \cdots + s_{n-1} x + s_n,$$

where $s_0 = 1$ and for each $k = 1, 2, \ldots, n$

$$s_k = s_k(a_1, a_2, \ldots, a_n) = \sum_{1 \leq p_1 < \cdots < p_k \leq n} \prod_{q=1}^{k} a_{p_q}.$$

We refer to s_k, depending on $a_1, a_2, \ldots, a_n$, as the kth *elementary symmetric function* of $a_1, a_2, \ldots, a_n$.

One may expand $p(x) = (x + a_1)(x + a_2)(x + a_3)$ as an example.

Theorem 4.11 *Suppose that $a_1, a_2, \ldots, a_n$ are distinct. Then*

$$(V_n(a_1, a_2, \ldots, a_n))^{-1} = (\alpha_{ij}),$$

where for each pair of i and j

$$\alpha_{ij} = \frac{(-1)^{1+j} \sum_{p_1 < \cdots < p_{n-j}} \prod_{q=1,\, p_q \neq i}^{n-j} a_{p_q}}{\prod_{k=1,\, k \neq i}^{n}(a_k - a_i)}.$$

PROOF. Recall from elementary linear algebra (Section 1.2) that the entries of the inverse of the matrix V are the cofactors of order $n-1$ divided by $\det V$, that is,

$$V^{-1} = \left(\frac{1}{\det V} c_{ij}\right)^T,$$

where c_{ij} is the cofactor of the (i,j)-entry of V.

In what follows we compute the cofactors c_{ij}. Let V_k be the matrix obtained from V by deleting row $k+1$ (the kth powers) and adjoining as a new nth row the nth powers of the a_i. We show

$$\det V_k = s_{n-k} \det V. \tag{4.9}$$

Augment V with the nth powers of the a_i as the $(n+1)$th row and with $(1, -x, (-x)^2, \ldots, (-x)^n)$ as the first column. Denote the resulting matrix by W. Then W is a Vandermonde matrix and

$$\begin{aligned}\det W &= (x+a_1) \cdots (x+a_n) \det V \\ &= (x^n + s_1 x^{n-1} + \cdots + s_{n-1} x + s_n) \det V. \end{aligned} \tag{4.10}$$

Expanding $\det W$ along the first column, we have

$$\det W = \det V_0 + x \det V_1 + \cdots + x^n \det V. \tag{4.11}$$

Identity (4.9) follows by comparing (4.10) and (4.11).

Now notice that each cofactor c_{ij} is a determinant of order $n-1$ in the same form as $\det V_k$. Let $V(\widehat{a_j})$ and $s_k(\widehat{a_j})$ denote, respectively, the $(n-1)$-square Vandermonde matrix and the kth elementary symmetric function of $a_1, a_2, \ldots, a_n$ without a_j. Using (4.9) we have

$$\begin{aligned} c_{ij} &= (-1)^{i+j} \det V(i|j) \\ &= (-1)^{i+j} s_{(n-1)-(i-1)}(\widehat{a_j}) \det V(\widehat{a_j}) \\ &= (-1)^{i+j} s_{n-i}(\widehat{a_j}) \det V(\widehat{a_j}). \end{aligned}$$

Thus,

$$\begin{aligned} \frac{1}{\det V} c_{ij} &= \frac{(-1)^{i+j} s_{n-i}(\widehat{a_j}) \det V(\widehat{a_j})}{\prod_{t>s}(a_t - a_s)} \\ &= \frac{(-1)^{i+j} s_{n-i}(\widehat{a_j})}{\prod_{s<j}(a_j - a_s) \prod_{j<t}(a_t - a_j)} \\ &= \frac{(-1)^{i+1} s_{n-i}(\widehat{a_j})}{\prod_{k=1,\ k \neq j}^n (a_k - a_j)} \\ &= \frac{(-1)^{i+1} \sum_{p_1 < \cdots < p_{n-i}} \prod_{q=1,\ p_q \neq j}^{n-i} a_{p_q}}{\prod_{k=1,\ k \neq j}^n (a_k - a_j)} \end{aligned}$$

or

$$\alpha_{ij} = \frac{(-1)^{1+j} \sum_{p_1 < \cdots < p_{n-j}} \prod_{q=1,\ p_q \neq i}^{n-j} a_{p_q}}{\prod_{k=1,\ k \neq i}^n (a_k - a_i)}. \quad \blacksquare$$

Problems

1. Find the solution to the equation in x

$$\begin{vmatrix} x^2 & 4 & 9 \\ x & 2 & 3 \\ 1 & 1 & 1 \end{vmatrix} = 0.$$

2. Evaluate the determinant

$$\begin{vmatrix} 1 & ax & a^2 + x^2 \\ 1 & ay & a^2 + y^2 \\ 1 & az & a^2 + z^2 \end{vmatrix}.$$

3. Find all solutions to the system of equations in $x_1, \ldots, x_n$

$$\begin{cases} x_1 + \cdots + x_n = a \\ x_1^2 + \cdots + x_n^2 = a^2 \\ \vdots \\ x_1^n + \cdots + x_n^n = a^n. \end{cases}$$

4. Let $x_1, x_2, \ldots, x_n$ be different numbers. Show that for any set of n numbers $y_1, y_2, \ldots, y_n$, there exists a polynomial $f(x)$ of degree at most $n-1$ such that

$$f(x_i) = y_i, \quad i = 1, 2, \ldots, n.$$

In particular, for any numbers $\lambda_1, \lambda_2, \ldots \lambda_n$, there exist polynomials $g(x)$ and $h(x)$ if each $\lambda_i \geq 0$ of degree at most $n-1$ such that

$$g(\lambda_i) = \overline{\lambda_i}, \quad h(\lambda_i) = \sqrt{\lambda_i}, \quad i = 1, 2, \ldots, n.$$

5. Let $A = \text{diag}(\lambda_1, \lambda_2, \ldots, \lambda_n)$, where $\lambda_i \neq \lambda_j$ for $i \neq j$. Show that for every normal matrix $B \in \mathbb{M}_n$, there exist a unitary matrix U and a polynomial f such that

$$B = U^* f(A) U.$$

6. Let $U = (u_{ij})$ be the p-square unitary matrix with

$$u_{ij} = \omega^{(i-1)(j-1)}, \quad i, j = 1, 2, \ldots, p,$$

where p is a prime integer and ω is a pth primitive root of unity. Show that all square submatrices of U are nonsingular.

7. Show that two n-square complex matrices A and B have the same set of eigenvalues if and only if

$$\text{tr } A^k = \text{tr } B^k, \quad k = 1, 2, \ldots, n.$$

8. Find the inverse, if it exists, of the Vandermonde matrix

$$\begin{pmatrix} 1 & 1 & 1 \\ 1 & x & y \\ 1 & x^2 & y^2 \end{pmatrix}.$$

9. Expand and find elementary symmetric functions for

$$p(x) = (x + a_1)(x + a_2)(x + a_3)(x + a_4).$$

10. Let W be the matrix obtained from $V = V_n(a_1, \ldots, a_n)$ by replacing the last row $(a_1^{n-1}, \ldots, a_n^{n-1})$ with $(a_1^n, \ldots, a_n^n)$. Show directly without using (4.9) that

$$\det W = (a_1 + \cdots + a_n) \det V.$$

11. Let $f_1(x)$, $f_2(x)$, $\ldots$, $f_n(x)$ be polynomials with degree at most $n-2$. Show that for any numbers $a_1, a_2, \ldots, a_n$

$$\begin{vmatrix} f_1(a_1) & f_1(a_2) & \cdots & f_1(a_n) \\ f_2(a_1) & f_2(a_2) & \cdots & f_2(a_n) \\ \vdots & \vdots & \vdots & \vdots \\ f_n(a_1) & f_n(a_2) & \cdots & f_n(a_n) \end{vmatrix} = 0.$$

12. Let $A = (a_{ij}) \in \mathbb{M}_n$ and $f_i(x) = a_{1i} + a_{2i}x + \cdots + a_{ni}x^{n-1}$ for $i = 1, 2, \ldots, n$. Show that

$$\begin{vmatrix} f_1(x_1) & f_1(x_2) & \cdots & f_1(x_n) \\ f_2(x_1) & f_2(x_2) & \cdots & f_2(x_n) \\ \vdots & \vdots & \vdots & \vdots \\ f_n(x_1) & f_n(x_2) & \cdots & f_n(x_n) \end{vmatrix} = \det A \prod_{1 \le i < j \le n} (x_j - x_i).$$

13. Let $a_1, a_2, \ldots, a_n$ be complex numbers. Show that

$$\begin{vmatrix} 1 & \cdots & 1 \\ a_1 & \cdots & a_n \\ \vdots & \vdots & \vdots \\ a_1^{n-2} & \cdots & a_n^{n-2} \\ a_2 a_3 \cdots a_n & \cdots & a_1 a_2 \cdots a_{n-1} \end{vmatrix} = (-1)^{n-1} \det V_n(a_1, \ldots, a_n).$$

14. Let A be an $n \times n$ matrix with eigenvalues $\lambda_1, \ldots, \lambda_n$. Show that

$$\det(\lambda I - A) = \lambda^n - \sigma_1 \lambda^{n-1} + \sigma_2 \lambda^{n-2} - \cdots + (-1)^n \sigma_n,$$

where $\sigma_k = s_k(\lambda_1, \lambda_2, \ldots, \lambda_n)$, $k = 1, 2, \ldots, n$. Describe σ_k in terms of principal minors (see Problem 17 on page 20.)

4.5 Hadamard Matrices

An n-square matrix A is called a *Hadamard matrix* if each entry of A is 1 or -1 and if the rows or columns of A are orthogonal, that is,

$$AA^T = nI \quad \text{or} \quad A^T A = nI.$$

Note that $AA^T = nI$ and $A^T A = nI$ are equivalent (Problem 5).

The following are two examples of Hadamard matrices:

$$\begin{pmatrix} 1 & 1 \\ 1 & -1 \end{pmatrix}, \quad \begin{pmatrix} 1 & 1 & 1 & 1 \\ 1 & 1 & -1 & -1 \\ 1 & -1 & -1 & 1 \\ 1 & -1 & 1 & -1 \end{pmatrix}.$$

Notice that if A is a Hadamard matrix, then so is AP for any matrix P with entries ± 1 satisfying $PP^T = I$. Thus, one may change the -1 in the first row of A to $+1$ by multiplying an appropriate matrix P with diagonal entries ± 1. There is only one 2×2 Hadamard matrix of this kind. Can one construct a 3×3 Hadamard matrix?

Theorem 4.12 *Let $n > 2$. A necessary condition for an n-square matrix A to be a Hadamard matrix is that n is a multiple of 4.*

PROOF 1. Let $A = (a_{ij})$ be an n-square Hadamard matrix. Then the equation $AA^T = nI$, noting that the entries of A are ± 1, yields

$$\sum_{k=1}^{n} a_{ik} a_{jk} = \begin{cases} 0, & \text{if } i \neq j, \\ n, & \text{if } i = j. \end{cases}$$

Upon computation, we have

$$\begin{aligned} \sum_{k=1}^{n}(a_{1k} + a_{2k})(a_{1k} + a_{3k}) &= \sum_{k=1}^{n} a_{1k}^2 + \sum_{k=1}^{n} a_{1k} a_{2k} \\ &\quad + \sum_{k=1}^{n} a_{1k} a_{3k} + \sum_{k=1}^{n} a_{2k} a_{3k} \\ &= \sum_{k=1}^{n} a_{1k}^2 \\ &= n. \end{aligned}$$

Observe that the possible values for $a_{1k}+a_{2k}$ and $a_{1k}+a_{3k}$ are $+2$, 0, and -2. Thus, each term in the summation

$$\sum_{k=1}^{n}(a_{1k}+a_{2k})(a_{1k}+a_{3k})$$

must be $+4$, 0, or -4. It follows that n is divisible by 4.

PROOF 2. Let P be an n-square matrix with diagonal entries 1 or -1 such that the first row of AP consists entirely of $+1$. Note that AP is also a Hadamard matrix. Since the second and third rows of AP are orthogonal to the first row, they must each have r $+1$ and r -1, and $n = 2r$ is an even number.

Let n_{-}^{+} be the number of columns of AP that contain a $+1$ of row 2 and a -1 of row 3. Similarly, define n_{+}^{-}, n_{+}^{+}, and n_{-}^{-}. Then

$$n_{+}^{+}+n_{-}^{+}=n_{+}^{+}+n_{+}^{-}=n_{-}^{-}+n_{-}^{+}=r.$$

Thus,

$$n_{+}^{+}=n_{-}^{-}, \quad n_{-}^{+}=n_{+}^{-}.$$

The orthogonality of rows 2 and 3 implies that

$$n_{+}^{+}+n_{-}^{-}=n_{+}^{-}+n_{-}^{+}.$$

This gives $n_{+}^{+}=n_{-}^{+}$. Therefore, $n = 2r = 4n_{+}^{+}$ is a multiple of 4. ∎

It has been conjectured that a Hadamard matrix of size $4k \times 4k$ exists for every positive integer k. The conjecture is not resolved yet.

The following theorem, verified by a direct computation, gives a way to construct Hadamard matrices of larger dimension.

Theorem 4.13 *If A is a Hadamard matrix, so is*

$$\begin{pmatrix} A & A \\ A & -A \end{pmatrix}. \quad (4.12)$$

By this theorem, Hadamard matrices H_n of order 2^n can be generated recursively by defining

$$H_1 = \begin{pmatrix} 1 & 1 \\ 1 & -1 \end{pmatrix}, \quad H_n = \begin{pmatrix} H_{n-1} & H_{n-1} \\ H_{n-1} & -H_{n-1} \end{pmatrix}, \quad n \geq 2. \quad (4.13)$$

Let

$$x_1 = \begin{pmatrix} 1 \\ -1+\sqrt{2} \end{pmatrix} \quad \text{and} \quad x_n = \begin{pmatrix} x_{n-1} \\ (-1+\sqrt{2})x_{n-1} \end{pmatrix}.$$

By a simple computation, H_1 has two eigenvalues $\pm\sqrt{2}$, and x_1 is an eigenvector corresponding to $\sqrt{2}$. This generalizes as follows.

Theorem 4.14 *Let H_n be defined as in (4.13). Then H_n has eigenvalues $+2^{\frac{n}{2}}$ and $-2^{\frac{n}{2}}$ each of multiplicity 2^{n-1}, and an eigenvector x_n corresponding to the positive eigenvalue $2^{\frac{n}{2}}$.*

PROOF. The proof is done by induction on n. The case of $n=1$ was discussed just prior to the theorem. Now for $n \geq 2$, we have

$$\begin{aligned}
\det(\lambda I - H_n) &= \begin{vmatrix} \lambda I - H_{n-1} & -H_{n-1} \\ -H_{n-1} & \lambda I + H_{n-1} \end{vmatrix} \\
&= \det\left((\lambda I - H_{n-1})(\lambda I + H_{n-1}) - H_{n-1}^2\right) \\
&= \det(\lambda^2 I - 2H_{n-1}^2) \\
&= \det(\lambda I - \sqrt{2}H_{n-1})\det(\lambda I + \sqrt{2}H_{n-1}).
\end{aligned}$$

This shows that each eigenvalue μ of H_{n-1} generates two eigenvalues $\pm\sqrt{2}\mu$ of H_n. The assertion then follows by the induction hypothesis, for H_{n-1} has eigenvalues $+2^{\frac{n-1}{2}}$ and $-2^{\frac{n-1}{2}}$ each of multiplicity 2^{n-2}.

To see the eigenvector part, we observe that, by induction again,

$$\begin{aligned}
H_n x_n &= \begin{pmatrix} H_{n-1} & H_{n-1} \\ H_{n-1} & -H_{n-1} \end{pmatrix} \begin{pmatrix} x_{n-1} \\ (-1+\sqrt{2})x_{n-1} \end{pmatrix} \\
&= \begin{pmatrix} \sqrt{2}H_{n-1}x_{n-1} \\ (2-\sqrt{2})H_{n-1}x_{n-1} \end{pmatrix} \\
&= 2^{\frac{n}{2}} \begin{pmatrix} x_{n-1} \\ (-1+\sqrt{2})x_{n-1} \end{pmatrix} \\
&= 2^{\frac{n}{2}} x_n. \blacksquare
\end{aligned}$$

Let J_n denote the n-square matrix whose entries are all equal to 1. We give a lower bound for the size of a Hadamard matrix that contains a J_n as a submatrix.

Hadamard Matrices

Theorem 4.15 *If A is an m-square Hadamard matrix that contains a J_n as a submatrix, then $m \geq n^2$.*

PROOF. We may assume by permutation that A is partitioned as

$$A = \begin{pmatrix} J_n & X \\ Y & Z_s \end{pmatrix},$$

where Z_s is an s-square matrix of entries ± 1, and $s = m - n$.
Since A is a Hadamard matrix of size $m = n + s$, we have

$$AA^T = (n+s)I_m,$$

which implies, by using the block form of A, that

$$J_n^2 + XX^T = (n+s)I_n.$$

Thus,
$$XX^T = (n+s)I_n - nJ_n. \tag{4.14}$$

The eigenvalues of the right-hand matrix in (4.14) are

$$n + s - n^2, \quad n+s, \quad \ldots, \quad n+s.$$

However, XX^T is positive semidefinite, and thus has nonnegative eigenvalues. Therefore, $n + s - n^2 \geq 0$ or $m \geq n^2$. ∎

Problems

1. Show that A is a Hadamard matrix, and then find A^4, where

$$A = \begin{pmatrix} 1 & 1 & 1 & -1 \\ 1 & 1 & -1 & 1 \\ 1 & -1 & 1 & 1 \\ -1 & 1 & 1 & 1 \end{pmatrix}.$$

2. Does there exist a 3×3 Hadamard matrix? How about a 6×6? Construct an 8×8 Hadamard matrix.

3. What is the determinant of an n-square Hadamard matrix?

4. Let $A = (a_{ij})$ be a 3×3 matrix. Show that if each $a_{ij} = 1$ or -1, then $\det A$ is an even number. What is the maximum for $\det A$?

5. Let A be an n-square matrix with entries ± 1. Show that
$$AA^T = nI \quad \Leftrightarrow \quad A^T A = nI.$$
Conclude that if A is Hadamard, then $\frac{1}{\sqrt{n}}A$ is orthogonal.

6. Find the eigenvalues and eigenvectors of the Hadamard matrix
$$\begin{pmatrix} 1 & 1 & 1 & 1 \\ 1 & -1 & 1 & -1 \\ 1 & 1 & -1 & -1 \\ 1 & -1 & -1 & 1 \end{pmatrix}.$$

7. Let $n \geq 2$ and define recursively, as in (4.13),
$$H_1 = \begin{pmatrix} 1 & 1 \\ 1 & -1 \end{pmatrix}, \quad H_n = \begin{pmatrix} H_{n-1} & H_{n-1} \\ H_{n-1} & -H_{n-1} \end{pmatrix},$$
and let
$$F_1 = \frac{1}{2}(I + 2^{-\frac{n}{2}} H_n), \quad F_2 = -\frac{1}{2}(I - 2^{-\frac{n}{2}} H_n).$$
Show that F_1 and F_2 are idempotent matrices and that
$$F_1 + F_2 = 2^{-\frac{n}{2}} H_n.$$

8. Let $n \geq 2$ and define recursively
$$E_1 = \begin{pmatrix} 0 & -1 \\ 1 & 0 \end{pmatrix}, \quad E_{n+1} = \begin{pmatrix} 0 & -E_n \\ E_n & 0 \end{pmatrix}.$$
Show that
   (a) $E_n^2 = (-1)^n I_{2^n}$,
   (b) E_n is symmetric if n is even, and skew-symmetric if n is odd,
   (c) $E_n H_n = (-1)^n H_n E_n$, where H_n is defined as in (4.13).

9. Let A be a square matrix with entries 1, -1, or 0. If each row and column of A contains only one nonzero entry 1 or -1, show that $A^k = I$ for some positive integer k.

10. How many $n \times n$ matrices of 0 and 1 entries are there for which the number of 1 in each row and column is even? (Answer: $2^{(n-1)(n-1)}$.)

4.6 Permutation and Doubly Stochastic Matrices

A square matrix P is called a *permutation matrix* if each row and column of P has exactly one 1 and all other entries are 0.

It is easy to see that there are $n!$ permutation matrices of size n. Furthermore, the product of two permutation matrices of the same size is a permutation matrix, and if P is a permutation matrix, then P is invertible, and $P^{-1} = P^T$ (Problem 1).

Our goal of this section is to show that every permutation matrix is a direct sum of primary permutation matrices under permutation similarity and that every doubly stochastic matrix is a convex combination of permutation matrices.

A matrix A of order n is said to be *reducible* if there exists a permutation matrix P such that

$$P^T A P = \begin{pmatrix} B & C \\ 0 & D \end{pmatrix}, \qquad (4.15)$$

where B and D are square matrices of order at least 1.

A matrix is said to be *irreducible* if it is not reducible. Note that a matrix of order 1 is irreducible. The matrix $P^T A P = P^{-1} A P$ in (4.15) is similar to A through the permutation matrix P. We say that they are *permutation similar*.

It is obvious that the diagonal entries of irreducible permutation matrices are all equal to 0, but not vice versa. For example,

$$\begin{pmatrix} 0 & 1 & 0 & 0 \\ 1 & 0 & 0 & 0 \\ 0 & 0 & 0 & 1 \\ 0 & 0 & 1 & 0 \end{pmatrix}.$$

Theorem 4.16 *Every reducible permutation matrix is permutation similar to a direct sum of irreducible permutation matrices.*

PROOF. Let A be an n-square reducible permutation matrix, as in (4.15). The matrix C in this case must be zero, for otherwise, let B

be $r \times r$ and D be $s \times s$, where $r + s = n$. Then B contains r 1 (in columns) and D contains s 1 (in rows). If C contained a 1, then A would have at least $r+s+1 = n+1$ 1, a contradiction. The assertion then follows by the induction on B and D. ∎

We now show that every n-square irreducible permutation matrix is permutation similar to the $n \times n$ primary permutation matrix

$$P = \begin{pmatrix} 0 & 1 & 0 & \cdots & 0 \\ 0 & 0 & 1 & \cdots & 0 \\ \vdots & \vdots & \vdots & \vdots & \vdots \\ 0 & 0 & 0 & \cdots & 1 \\ 1 & 0 & 0 & \cdots & 0 \end{pmatrix}. \tag{4.16}$$

Theorem 4.17 *A primary permutation matrix is irreducible.*

PROOF. Suppose the $n \times n$ primary permutation matrix P is reducible. Let $S^T P S = J_1 \oplus \cdots \oplus J_k$, $k \geq 2$, where S is some permutation matrix and the J_i are irreducible matrices with order $< n$.

The rank of $P - I$ is $n - 1$, for $\det(P - I) = 0$ and the submatrix of size $n - 1$ by deleting the last row and the last column from $P - I$ is nonsingular. It follows that

$$\operatorname{rank}(S^T P S - I) = \operatorname{rank}(S^T (P - I) S) = n - 1.$$

By using the above decomposition, we have

$$\operatorname{rank}(S^T P S - I) = \sum_{i=1}^{k} \operatorname{rank}(J_i - I) \leq n - k < n - 1.$$

This is a contradiction. The proof is complete. ∎

The eigenvalues of the $n \times n$ primary permutation matrix P are exactly all the roots of the equation $\lambda^n = 1$, that is, $1, \omega, \ldots, \omega^{n-1}$, where ω is an nth primitive root of unity, since

$$\det(\lambda I - P) = \lambda^n - 1$$

by a direct computation. In addition, for any positive integer $k < n$,

$$P^{n-1} = P^T, \quad P^n = I_n, \quad P^k = \begin{pmatrix} 0 & I_{n-k} \\ I_k & 0 \end{pmatrix}.$$

Theorem 4.18 *A permutation matrix is irreducible if and only if it is permutation similar to a primary permutation matrix.*

PROOF. Let Q be an $n \times n$ permutation matrix and P the $n \times n$ primary permutation matrix in (4.16). If Q is permutation similar to P, then Q is irreducible by the previous theorem. Conversely, suppose that Q is irreducible. We show that Q can be brought to P through simultaneous row and column permutations.

Let the 1 of the first row be in the position $(1, i_1)$. Then $i_1 \neq 1$ since Q is irreducible. If $i_1 = 2$, we proceed to the next step, considering the 1 in the second row. Otherwise, $i_1 > 2$. Permute columns 2 and i_1 so that the 1 is placed in the $(1, 2)$-position.

Permute rows 2 and i_1 to get a matrix Q_1. This matrix is permutation similar to Q and also irreducible. If the $(2, 3)$-entry of Q_1 is 1, we go on to next step. Otherwise, let the $(2, i_2)$-entry be 1, $i_2 \neq 3$. If $i_2 = 1$, then Q_1 would be reducible, for all entries in the first two columns but not in the first two rows equal 0. Thus, $i_2 \geq 3$. Permute columns 3 and i_2 so that the 1 is in the $(2, 3)$-position.

Note that the 1 in the $(1, 2)$-position was not affected by the permutations in the second step. Continuing in this way, one obtains the permutation matrix P in the form of (4.16). Since the product of a sequence of permutation matrices is also a permutation matrix, we thus have a permutation matrix S such that

$$S^T Q S = S^{-1} Q S = P. \blacksquare$$

Combining the above theorems, we see that every reducible permutation matrix is permutation similar to a direct sum of primary permutation matrices. Moreover, the rank of an n-square irreducible permutation matrix minus I is $n - 1$ (Problem 3).

Theorem 4.19 *Let Q be an n-square permutation matrix. Then Q is irreducible if and only if the eigenvalues of Q are $1, \omega, \ldots, \omega^{n-1}$, where ω is an nth primitive root of unity.*

PROOF. If Q is irreducible, then Q is similar to the $n \times n$ primary permutation matrix, according to Theorem 4.18, which has the eigenvalues $1, \omega, \ldots, \omega^{n-1}$, so does matrix Q.

Conversely, suppose that $1, \omega, \ldots, \omega^{n-1}$ are the eigenvalues of Q. Note that $\omega^k \neq 1$ for any $1 \leq k < n$ since ω is an nth primitive root of unity. If Q is reducible, then we may write

$$S^T Q S = J_1 \oplus \cdots \oplus J_k,$$

where S is a permutation matrix, and the J_i are primary permutation matrices with order less than n.

The eigenvalues of those J_i are the eigenvalues of Q, none of which is an nth primitive root of unity, for the order of every J_i is less than n. This is a contradiction. Thus, Q is irreducible. ∎

We next present a beautiful relation between permutation matrices and doubly stochastic matrices, a type of matrices that plays an important role in statistics and in some other subjects.

A square matrix is said to be *doubly stochastic* if all entries of the matrix are nonnegative and the sum of the entries in each row and each column equals 1. Equivalently, a matrix A with nonnegative entries is doubly stochastic if

$$e^T A = e^T \quad \text{and} \quad Ae = e, \quad \text{where } e = (1, \ldots, 1)^T. \tag{4.17}$$

It is readily seen that permutation matrices are doubly stochastic and so is the product of two doubly stochastic matrices.

We show that a matrix is a doubly stochastic matrix if and only if it is a convex combination of finite permutation matrices. To prove this, we need a result, which is of interest in its own right.

Theorem 4.20 (Frobenius-König) *Let A be an n-square complex matrix. Then every product of n entries of A taken from distinct rows and columns equals 0, in symbols,*

$$a_{1i_1} a_{2i_2} \cdots a_{ni_n} = 0, \quad \{i_1, i_2, \ldots, i_n\} = \{1, 2, \ldots, n\}, \tag{4.18}$$

if and only if A contains an $r \times s$ zero submatrix, where $r + s = n + 1$.

PROOF. First notice that property (4.18) of A will remain true when row or column permutations are applied to A. In other words, an n-square matrix A has property (4.18) if and only if PAQ has the property, where P and Q are any n-square permutation matrices.

SEC. 4.6   PERMUTATION AND DOUBLY STOCHASTIC MATRICES   127

Sufficiency: We may assume by permutation that the $r \times s$ zero submatrix is in the lower left corner, and write

$$A = \begin{pmatrix} B & C \\ 0 & D \end{pmatrix}.$$

Since $n - r = s - 1$, B is of size $(s-1) \times s$. Thus, there must be a zero among any s entries taken from the first s columns and any s different rows. Therefore, every product $a_{1i_1} a_{2i_2} \cdots a_{ni_n}$ has to contain a zero factor, hence equals zero.

Necessity: If all the entries of A are zero, there is nothing to prove. Suppose A has a nonzero entry and consider the submatrix obtained from A by deleting the row and the column that contain the nonzero entry. An application of induction on the $(n-1) \times (n-1)$ submatrix results in a zero submatrix of size $p \times q$, where $p+q = (n-1)+1 = n$. We thus may write A, by permutation, as

$$A = \begin{pmatrix} B & C \\ 0 & D \end{pmatrix},$$

where B is $q \times q$ and D is $p \times p$. Since every product of the entries of A from different rows and columns is 0, this property must be inherited by B or D, say B. Applying the induction to B, we see that B has an $l \times s$ zero submatrix such that $l + s = q + 1$. Putting this zero submatrix in the lower-left corner of B, we see that A has an $r \times s$ zero submatrix, where $r = p + l$ and $r + s = n + 1$. ∎

Theorem 4.21 (Birkhoff) *A matrix A is doubly stochastic if and only if it is a convex combination of permutation matrices.*

PROOF. To show sufficiency, let A be a *convex combination* of permutation matrices $P_1, P_2, \ldots, P_m$, that is,

$$A = t_1 P_1 + t_2 P_2 + \cdots + t_m P_m,$$

where $t_1, t_2, \ldots, t_m$ are nonnegative numbers of sum equal to 1. Then it is easy to see that $e^T A = e^T$ and $Ae = e$, where $e = (1, \ldots, 1)^T$. By (4.17) A is doubly stochastic.

For necessity, we apply induction on the number of zero entries of the doubly stochastic matrices. If A has (at most) $n^2 - n$ zeros, then

A is a permutation matrix, and we have nothing to show. Suppose that the doubly stochastic matrices with at least k zeros are convex combinations of permutation matrices. We show that the assertion holds for the doubly stochastic matrices with $k-1$ zeros.

Let A be an n-square doubly stochastic matrix of $k-1$ zero entries. If every product of the entries A from distinct rows and columns is zero, then A may be written as, up to permutation,

$$A = \begin{pmatrix} B & C \\ 0 & D \end{pmatrix},$$

where the zero submatrix is of size $r \times s$ with $r + s = n + 1$.

Since the entries in each column A add up to 1, the sum of all entries of B equals s. Similarly, by considering rows, the sum of all entries of D is r. Thus, the sum of all entries of A would be at least $r + s = n + 1$. This is impossible, for the sum of all entries of A is n. Therefore, some product $a_{1i_1} a_{2i_2} \cdots a_{ni_n} \neq 0$.

Let P_1 be a permutation matrix with 1 in the positions (j, i_j), $j = 1, 2, \ldots, n$, and 0 elsewhere. Consider the matrix

$$E = (1-\delta)^{-1}(A - \delta P_1),$$

where $\delta = \min\{a_{1i_1}, a_{2i_2}, \ldots, a_{ni_n}\}$.

It is readily seen by (4.17) that E is also a doubly stochastic matrix and that E has at least one more zero than A. By the induction hypothesis, there are positive numbers $t_2, \ldots, t_m$ of sum 1, and permutations matrices $P_2, \ldots, P_m$, such that

$$E = t_2 P_2 + \cdots + t_m P_m.$$

It follows that

$$A = \delta P_1 + (1-\delta)t_2 P_2 + \cdots + (1-\delta)t_m P_m,$$

where P_i are permutation matrices, and their coefficients are nonnegative and sum up to 1. ∎

Problems

1. Show that the determinant of a permutation matrix is ± 1 and that permutation matrices are unitary and normal.

2. Find permutations that bring the reducible permutation matrix

$$P = \begin{pmatrix} 0 & 0 & 1 & 0 \\ 0 & 0 & 0 & 1 \\ 1 & 0 & 0 & 0 \\ 0 & 1 & 0 & 0 \end{pmatrix}$$

to a direct sum of irreducible matrices. Show that $P^2 = I$.

3. Let P be an $n \times n$ irreducible permutation matrix. Show that

$$\operatorname{rank}(P - I) = n - 1.$$

4. Let P be an $n \times n$ irreducible permutation matrix. Show that P^k is irreducible if and only if $(n, k) = 1$, that is, n and k have no common positive divisors other than 1.

5. Let P be an n-square permutation matrix. Show that P is irreducible if and only if P has the property $P^m = I_n \Leftrightarrow n|m$.

6. Let P be an $n \times n$ irreducible permutation matrix. Show that P is diagonalizable over $\mathbb{C}$ but not over $\mathbb{R}$ when $n \geq 3$.

7. Show that if two permutation matrices are similar, then they are permutation similar.

8. Show that for any $n \times n$ permutation matrix P, $P^{n!} = I$, and further that if P is irreducible, then $P^n = I$. Is the converse true?

9. Prove or disprove that a symmetric permutation matrix (of odd or even order greater than 1) is reducible.

10. Find the rank of the partitioned permutation matrix Q, where

$$Q = \begin{pmatrix} 0 & 0 & I & 0 & 0 \\ 0 & 0 & 0 & I & 0 \\ 0 & 0 & 0 & 0 & I \\ I & 0 & 0 & 0 & 0 \\ 0 & I & 0 & 0 & 0 \end{pmatrix}.$$

11. Show that an n-square matrix A with nonnegative entries is reducible if and only if there exists a proper subset $\{i_1, \ldots, i_k\}$ of $\{1, \ldots, n\}$ such that

$$\operatorname{Span}\{Ae_{i_1}, \ldots, Ae_{i_k}\} \subseteq \operatorname{Span}\{e_{i_1}, \ldots, e_{i_k}\},$$

where the e_i are the standard basis vectors for $\mathbb{C}^n$.

12. Show that the product of two doubly stochastic matrices is a doubly stochastic matrix. How about the sum?

13. Show that if $U = (u_{ij})$ is a unitary matrix, then $A = (|u_{ij}|^2)$ is a doubly stochastic matrix. How about $B = (|u_{ij}||v_{ij}|)$ where $V = (v_{ij})$ is a unitary matrix of the same size as U?

14. Show that the following determinant is zero:

$$\begin{vmatrix} a & b & 0 & 0 & 0 \\ c & d & 0 & 0 & 0 \\ e & f & 0 & 0 & 0 \\ g & h & i & j & k \\ l & m & n & o & p \end{vmatrix}.$$

15. Show that every $n \times n$ doubly stochastic matrix is a convex combination of at most $n^2 - 2n + 2$ permutation matrices.

16. A *permutation of an n-element set* $\{1, 2, \ldots, n\}$ is a mapping

$$p : 1 \to i_1,\ 2 \to i_2,\ \ldots,\ n \to i_n,$$

written as

$$p = \begin{pmatrix} 1 & 2 & \cdots & n \\ i_1 & i_2 & \cdots & i_n \end{pmatrix}.$$

Associate to such a p a permutation matrix $P = f(p)$, which is 1 in the (k, i_k)-position and 0 elsewhere, $k = 1, 2, \ldots, n$, and define the product of

$$p = \begin{pmatrix} 1 & 2 & \cdots & n \\ i_1 & i_2 & \cdots & i_n \end{pmatrix} \quad \text{and} \quad q = \begin{pmatrix} i_1 & i_2 & \cdots & i_n \\ j_1 & j_2 & \cdots & j_n \end{pmatrix}$$

by

$$pq = \begin{pmatrix} 1 & 2 & \cdots & n \\ j_1 & j_2 & \cdots & j_n \end{pmatrix}.$$

Show that

$$f(pq) = f(p)f(q).$$

17. Show that any permutation matrix can be expressed as the product of at most $n - 1$ symmetric permutation matrices.

CHAPTER 5

Unitary Matrices and Contractions

INTRODUCTION This chapter studies unitary matrices and contractions. Section 5.1 gives basic properties of unitary matrices, Section 5.2 discusses the structure of real orthogonal matrices under similarity, and Section 5.3 develops metric spaces and the fixed-point theorem of strict contractions. Section 5.4 deals with the connections of contractions with unitary matrices, Section 5.5 concerns the unitary similarity of real matrices, and Section 5.6 presents a trace inequality for unitary matrices, relating the average of the eigenvalues of each of two unitary matrices to that of their product.

5.1 Properties of Unitary Matrices

A *unitary matrix* is a square complex matrix satisfying
$$U^*U = UU^* = I.$$
Notice that $U^* = U^{-1}$ and $|\det U| = 1$ for any unitary matrix U. A complex (real) matrix A is called *complex (real) orthogonal* if
$$A^T A = AA^T = I.$$
Unitary matrices and complex orthogonal matrices are different in general. But real unitary and real orthogonal matrices are the same.

Theorem 5.1 *Let $U \in \mathbb{M}_n$ be a unitary matrix. Then*

1. *$\|Ux\| = \|x\|$ for every $x \in \mathbb{C}^n$,*

2. *$|\lambda| = 1$ for every eigenvalue λ of U,*

3. *$U = V \operatorname{diag}(\lambda_1, \ldots, \lambda_n) V^*$, where V is unitary and each $|\lambda_i| = 1$,*

4. *the column vectors of U form an orthonormal basis for $\mathbb{C}^n$.*

PROOF. (1) is obtained by rewriting the norm as an inner product:
$$\|Ux\| = \sqrt{(Ux, Ux)} = \sqrt{(x, U^*Ux)} = \sqrt{(x,x)} = \|x\|.$$

To show (2), let x be a unit eigenvector of U corresponding to eigenvalue λ. Then, by using (1),
$$|\lambda| = |\lambda|\|x\| = \|\lambda x\| = \|Ux\| = \|x\| = 1.$$

(3) is by the spectral decomposition theorem (Theorem 3.4).

For (4), suppose that u_i is the ith column of U, $i = 1, \ldots, n$. Then the matrix identity $U^*U = I$ is equivalent to
$$\begin{pmatrix} u_1^* \\ \vdots \\ u_n^* \end{pmatrix} (u_1, \ldots, u_n) = \begin{pmatrix} 1 & & 0 \\ & \ddots & \\ 0 & & 1 \end{pmatrix}.$$

It follows that $(u_i, u_j) = u_j^* u_i = 1$ if $i = j$, and 0 otherwise. ∎

An interesting observation follows: Note that for any unitary U,
$$\operatorname{adj}(U) = (\det U)U^{-1} = (\det U)U^*.$$
If we partition U as
$$U = \begin{pmatrix} u & \alpha \\ \beta & U_1 \end{pmatrix}, \quad u \in \mathbb{C},$$
then, by comparing the $(1, 1)$-entries in $\operatorname{adj}(U) = (\det U)U^*$,
$$\det U_1 = \det U \, \bar{u},$$
which is also a consequence of Theorem 2.3.

As we saw in Theorem 5.1, the eigenvalues of a unitary matrix are necessarily equal to 1 in absolute value. The converse is not true in general. We have, however, the following result.

Sec. 5.1   Properties of Unitary Matrices

Theorem 5.2 Let $A \in \mathbb{M}_n$ have all the eigenvalues equal to 1 in absolute value. Then A is unitary if, for all $x \in \mathbb{C}^n$,

$$\|Ax\| \leq \|x\|.$$

PROOF 1. The given inequality is equivalent to

$$\|Ax\| \leq 1, \quad \text{for all unit } x \in \mathbb{C}^n.$$

This specifies that $\sigma_{\max}(A) \leq 1$ (Problem 7, Section 3.6).

On the other hand, the identity $|\det A|^2 = \det(A^*A)$ implies that the product of eigenvalues in absolute value equals the product of singular values. If A has only eigenvalues 1 in absolute value, then the smallest singular value of A has to be 1; thus, all the singular values of A are equal to 1. Therefore, $A^*A = I$, and A is unitary.

PROOF 2. Let $A = U^*DU$ be a Schur decomposition of A, where U is a unitary matrix, and D is an upper-triangular matrix:

$$D = \begin{pmatrix} \lambda_1 & t_{12} & \cdots & t_{1n} \\ 0 & \lambda_2 & \ddots & \vdots \\ \vdots & \ddots & \ddots & t_{n-1,n} \\ 0 & \cdots & 0 & \lambda_n \end{pmatrix},$$

where $|\lambda_i| = 1$, $i = 1, 2, \ldots, n$, and t_{ij} are complex numbers.

Take $x = U^*e_n = U^*(0, \ldots, 0, 1)^T$. Then $\|x\| = 1$. We have

$$\|Ax\| = \|De_n\| = (|t_{1n}|^2 + \cdots + |t_{n-1,n}|^2 + |\lambda_n|^2)^{\frac{1}{2}}.$$

The conditions $\|Ax\| \leq 1$ for unit x and $|\lambda_n| = 1$ force each $t_{in} = 0$ for $i = 1, 2, \ldots, n-1$. By induction, one sees that D is a diagonal matrix with the λ_i on the diagonal. Thus, A is unitary, for

$$A^*A = U^*D^*UU^*DU = U^*D^*DU = U^*U = I. \blacksquare$$

We now show a result on the singular values of the principal submatrices of a unitary matrix.

Theorem 5.3 *Let U be a unitary matrix partitioned as*

$$U = \begin{pmatrix} A & B \\ C & D \end{pmatrix},$$

where A is $m \times m$ and D is $n \times n$. If $m = n$, then A and D have the same singular values. If $m < n$ and if the singular values of A are $\sigma_1, \ldots, \sigma_m$, then the singular values of D are $\sigma_1, \ldots, \sigma_m, \overbrace{1, \ldots, 1}^{n-m}$.

PROOF. Since U is unitary, the identities $U^*U = UU^* = I$ imply

$$A^*A + C^*C = I_m, \quad CC^* + DD^* = I_n.$$

It follows that

$$A^*A = I_m - C^*C, \quad DD^* = I_n - CC^*.$$

Note that CC^* and C^*C have the same nonzero eigenvalues. Hence, $I_n - CC^*$ and $I_m - C^*C$ have the same eigenvalues except $n - m$ 1, that is, A^*A and DD^* have the same eigenvalues except $n - m$ 1. Therefore, if $m = n$, then A and D have the same singular values, and if $m < n$ and A has singular values $\sigma_1, \ldots, \sigma_m$, then the singular values of D are $\sigma_1, \ldots, \sigma_m$, plus $n - m$ 1. ∎

An interesting result on the unitary matrix U in Theorem 5.3 is

$$|\det A| = |\det D|.$$

In other words, the complementary principal submatrices of a unitary matrix always have the same determinant in absolute value.

Problems

1. Which of the items in Theorem 5.1 implies that U is a unitary matrix?
2. Show that for any $\theta_1, \ldots, \theta_n \in \mathbb{R}$, $\text{diag}(e^{i\theta_1}, \ldots, e^{i\theta_n})$ is unitary.
3. What are the singular values of a unitary matrix?
4. Find an $m \times n$ matrix V, $m \neq n$, such that $V^*V = I_n$.
5. Let $A = \begin{pmatrix} 0 & x \\ 0 & 0 \end{pmatrix}$, where $x \in \mathbb{C}$. What are the eigenvalues and singular values of A in terms of x?

6. Show that for any $\theta \in \mathbb{R}$, the following two matrices are similar:

$$\begin{pmatrix} e^{i\theta} & 0 \\ 0 & e^{-i\theta} \end{pmatrix} \quad \text{and} \quad \begin{pmatrix} \cos\theta & \sin\theta \\ -\sin\theta & \cos\theta \end{pmatrix}.$$

Note: They are both unitary; the first is complex, the second is real.

7. Show that any 2×2 unitary matrix with determinant equal to 1 is similar to a real orthogonal matrix.

8. Let A and C be m- and n-square matrices, respectively, and let

$$M = \begin{pmatrix} A & B \\ 0 & C \end{pmatrix}.$$

Show that M is unitary if and only if $B = 0$ and A and C are unitary.

9. Show that the 2×2 block matrix below is real orthogonal:

$$\begin{pmatrix} \sqrt{\lambda}\,I & -\sqrt{1-\lambda}\,I \\ \sqrt{1-\lambda}\,I & \sqrt{\lambda}\,I \end{pmatrix}, \quad \lambda \in [0,1].$$

10. If A is a unitary matrix with all eigenvalues real, show that

$$A^2 = I \quad \text{and} \quad A^* = A.$$

11. If A is similar to a unitary matrix, show that A^* is similar to A^{-1}.

12. Let $A \in \mathbb{M}_n$ be Hermitian. Show that $(A - iI)^{-1}(A + iI)$ is unitary.

13. Let A be an $n \times n$ unitary matrix. If $A - I$ is nonsingular, show that $i(A - I)^{-1}(A + I)$ is Hermitian.

14. Let a and b be real numbers such that $a^2 - b^2 = 1$, $ab \neq 0$, and let

$$K = \begin{pmatrix} a & bi \\ -bi & a \end{pmatrix}.$$

(a) Show that K is complex orthogonal but not unitary, that is,

$$K^T K = I \quad \text{but} \quad K^* K \neq I.$$

(b) Let $a = \sqrt{2}$ and $b = 1$. Find the eigenvalues of K.

(c) Let $a = \frac{1}{2}(e + e^{-1})$ and $b = \frac{1}{2}(e - e^{-1})$, where $e = 2.718\ldots$. Show that the eigenvalues of K are e and e^{-1}.

(d) Let $a = \frac{1}{2}(e^t + e^{-t})$ and $b = \frac{1}{2}(e^t - e^{-t})$, $t \in \mathbb{R}$. What are the eigenvalues of K? Is the trace of K bounded?

15. Let A be an n-square complex matrix. Show that $A = 0$ if A satisfies
$$|(Ay, y)| \leq \|y\|, \quad \text{for all } y \in \mathbb{C}^n,$$
or
$$|(Ay, y)| \leq \|Ay\|, \quad \text{for all } y \in \mathbb{C}^n.$$
(Hint: If $A \neq 0$, then $(Ay_0, y_0) \neq 0$ for some $y_0 \in \mathbb{C}^n$.)

16. Let $A \in \mathbb{M}_n$ and $\alpha \in (0, 1)$. What can be said about A if
$$|(Ay, y)| \leq (y, y)^\alpha, \quad \text{for all } y \in \mathbb{C}^n?$$

17. Let $A \in \mathbb{M}_n$ have all eigenvalues equal to 1 in absolute value. Show that A is unitary if A satisfies
$$|(Ay, y)| \leq \|Ay\|^2, \quad \text{for all } y \in \mathbb{C}^n,$$
or
$$|(Ay, y)| \leq \|y\|^2, \quad \text{for all } y \in \mathbb{C}^n.$$

18. Let A be an $n \times n$ complex matrix having the largest and the smallest singular values $\sigma_{\max}(A)$ and $\sigma_{\min}(A)$, respectively. Show that
$$(Ay, Ay) \leq (y, y), \quad \text{for all } y \in \mathbb{C}^n, \quad \Rightarrow \quad \sigma_{\max}(A) \leq 1$$
and
$$(y, y) \leq (Ay, Ay), \quad \text{for all } y \in \mathbb{C}^n, \quad \Rightarrow \quad \sigma_{\min}(A) \geq 1.$$

19. Let $A \in \mathbb{M}_n$ have all eigenvalues equal to 1 in absolute value. Show that A is unitary if A satisfies, for some real $\alpha \neq \frac{1}{2}$,
$$|(Ay, y)| \leq (Ay, Ay)^\alpha, \quad \text{for all unit } y \in \mathbb{C}^n.$$

(Hint: Assume that A is upper-triangular with $a_{11} = 1$, $a_{1i} > 0$ for some $i > 1$. Take $y = (\cos t, 0, \ldots, 0, \sin t, 0, \ldots, 0)$ and consider the behavior of the function $f(t) = (Ay, Ay)^\alpha - |(Ay, y)|$ at 0.)

5.2 Real Orthogonal Matrices

This section is devoted to real orthogonal matrices, the real matrices A satisfying $AA^T = A^T A = I$. We discuss the structure of real orthogonal matrices under similarity and show that real orthogonal matrices with a commutativity condition are necessarily involutions.

We begin with 2×2 real orthogonal matrices A:

$$A = \begin{pmatrix} a & b \\ c & d \end{pmatrix}, \quad a, b, c, d \in \mathbb{R}.$$

The identities $AA^T = A^T A = I$ imply several equations in a, b, c, and d, one of which is $a^2 + b^2 = 1$. Since a is real between -1 and 1, one may set $a = \cos\theta$ for some $\theta \in \mathbb{R}$, then get b, c, and d in terms of θ. Thus, there are only two types of 2×2 real orthogonal matrices:

$$\begin{pmatrix} \cos\theta & \sin\theta \\ -\sin\theta & \cos\theta \end{pmatrix}, \quad \begin{pmatrix} \cos\theta & \sin\theta \\ \sin\theta & -\cos\theta \end{pmatrix}, \quad \theta \in \mathbb{R}, \tag{5.1}$$

where the first type is called *rotation* and the second *reflection*.

We show that a real orthogonal matrix is similar to a direct sum of real orthogonal matrices of order 1 or 2.

Theorem 5.4 *Every real orthogonal matrix is real orthogonally similar to a direct sum of real orthogonal matrices of order at most 2.*

PROOF. Let A be an $n \times n$ real orthogonal matrix. We apply induction on n. If $n = 1$ or 2, there is nothing to prove.

Suppose $n > 2$. If A has a real eigenvalue λ with a real unit eigenvector x, then

$$Ax = \lambda x, \quad x \neq 0 \quad \Rightarrow \quad x^T A^T Ax = \lambda^2 x^T x.$$

Thus, $\lambda = \pm 1$, say, 1. Extend the real unit eigenvector x to a real orthogonal matrix P. Then $P^T AP$ has $(1, 1)$-entry 1 and 0 elsewhere in the first column. Write in symbols

$$P^T AP = \begin{pmatrix} 1 & u \\ 0 & A_1 \end{pmatrix}.$$

The orthogonality of A implies $u = 0$. Notice that A_1 is also real orthogonal. The conclusion then follows from an induction on A_1.

Assume that A has no real eigenvalues. Then for any nonzero $x \in \mathbb{R}^n$, vectors x and Ax cannot be linearly dependent. Recall the angle $\angle_{x,y}$ between two vectors x and y and note that $\angle_{x,y} = \angle_{Ax,Ay}$ due to the orthogonality of A. Define $f(x)$ to be the angle function

$$f(x) = \angle_{x,Ax} = \cos^{-1} \frac{(x, Ax)}{\|x\| \|Ax\|}.$$

Then $f(x)$ is continuous on the compact set $S = \{x \in \mathbb{R}^n : \|x\| = 1\}$.

Let $\theta_0 = \angle_{x_0, Ax_0}$ be the minimum of $f(x)$ on S. Let y_0 be the unit vector in $\mathrm{Span}\{x_0, Ax_0\}$ such that $\angle_{x_0, y_0} = \angle_{y_0, Ax_0}$. Then by Theorem 1.9, we have

$$\theta_0 \leq \angle_{y_0, Ay_0} \leq \angle_{y_0, Ax_0} + \angle_{Ax_0, Ay_0} = \frac{\theta_0}{2} + \frac{\theta_0}{2} = \theta_0$$

and $Ax_0 \in \mathrm{Span}\{y_0, Ay_0\}$. Thus, Ay_0 has to belong to $\mathrm{Span}\{x_0, y_0\}$. It follows, since Ax_0 is also in the subspace, that $\mathrm{Span}\{x_0, y_0\}$ is an invariant subspace under A. We thus write A, up to similarity by taking a suitable orthonormal basis (equivalently, via a real orthogonal matrix), as

$$A = \begin{pmatrix} T_0 & 0 \\ 0 & B \end{pmatrix},$$

where T_0 is a 2×2 matrix, and B is a matrix of order $n - 2$. Since A is orthogonal, so are T_0 and B. An application of the induction hypothesis to B completes the proof. ∎

Another way to attack the problem is to consider the eigenvectors of A. One may again focus on the nonreal eigenvalues. Since A is real, the characteristic polynomial of A has real coefficients, and the nonreal eigenvalues of A thus occur in conjugate pairs. Furthermore, their eigenvectors are in the forms $\alpha + \beta i$ and $\alpha - \beta i$, where α and β are real, $(\alpha, \beta) = 0$, and $A\alpha = a\alpha - b\beta$, $A\beta = b\alpha + a\beta$ for some real a and b with $a^2 + b^2 = 1$. Matrix A will have the desired form (via orthogonal similarity) by choosing a suitable real orthonormal basis.

Our next theorem shows that a matrix with certain commuting property is necessarily an involution. For this purpose, we need a result that is of interest in its own right:

If two complex square matrices F and G of orders m and n, respectively, have no eigenvalues in common, then the matrix equation $FX - XG = 0$ has a unique solution $X = 0$.

To see this, rewrite the equation as $FX = XG$. Then for every positive integer k, $F^k X = XG^k$. It follows that

$$f(F)X = Xf(G)$$

for every polynomial f. In particular, we take f to be the characteristic polynomial $\det(\lambda I - F)$ of F; then $f(F) = 0$, and thus $Xf(G) = 0$. However, since F and G have no eigenvalues in common, $f(G)$ is nonsingular and hence $X = 0$.

Theorem 5.5 *Let A and U be real orthogonal matrices of the same size. If U has no repeated eigenvalues and if*

$$UA = AU^T,$$

then A is an involution, that is, $A^2 = I$.

PROOF. By the previous theorem, let P be a real orthogonal matrix such that $P^{-1}UP$ is a direct sum of orthogonal matrices V_i of order 1 or 2. The identity $UA = AU^T$ results in

$$(P^{-1}UP)(P^{-1}AP) = (P^{-1}AP)(P^{-1}UP)^T.$$

Partition $P^{-1}AP$ conformally with $P^{-1}UP$ as $P^{-1}AP = (B_{ij})$, where the B_{ij} are matrices whose number of rows (or columns) is 1 or 2. Then $UA = AU^T$ gives

$$V_i B_{ij} = B_{ij} V_j^T, \quad i,j = 1, \ldots, k. \tag{5.2}$$

Since U, thus $P^{-1}UP$, has no repeated eigenvalues, we have

$$B_{ij} = 0, \quad i \neq j.$$

Hence $P^{-1}AP$ is a direct sum of matrices of order no more than 2:

$$P^{-1}AP = B_{11} \oplus \cdots \oplus B_{kk}.$$

The orthogonality of A, thus $P^{-1}AP$, implies that each B_{ii} is either an orthogonal matrix of order 1 or an orthogonal matrix of form (5.1). Obviously $B_{ii}^2 = I$ if B_{ii} is ± 1 or a reflection. Now suppose B_{ii} is a rotation; then V_i is not a rotation. Otherwise, V_i and B_{ii} are both rotations and hence commute (Problem 4), so that $V_i^2 B_{ii} = B_{ii}$ and $V_i^2 = I$. Using the rotation in (5.1), we have $V_i = \pm I$, contradicting the fact that V_i has 2 distinct eigenvalues. Thus, V_i is a reflection, hence orthogonally similar to $\mathrm{diag}(1, -1)$. It follows that B_{ii} is similar to $\mathrm{diag}(\pm 1, \pm 1)$ by (5.2). In either case $B_{ii}^2 = I$. Thus, $(P^{-1}AP)^2 = I$ and $A^2 = I$. ■

Problems

1. Give a 2×2 matrix such that $A^2 = I$ but $A^*A \neq I$.

2. When is an upper-triangular matrix (complex or real) orthogonal?

3. If A is a 2×2 real matrix with a complex eigenvalue $\lambda = a + bi$, $a, b \in \mathbb{R}$, $b \neq 0$, show that A is similar to the real matrix
$$\begin{pmatrix} a & b \\ -b & a \end{pmatrix}.$$

4. Verify that A and B commute, that is, $AB = BA$, where
$$A = \begin{pmatrix} \cos\alpha & \sin\alpha \\ -\sin\alpha & \cos\alpha \end{pmatrix}, \quad B = \begin{pmatrix} \cos\beta & \sin\beta \\ -\sin\beta & \cos\beta \end{pmatrix}, \quad \alpha, \beta \in \mathbb{R}.$$

5. Show that a real matrix P is orthogonal projection if and only if P is orthogonally similar to a matrix in the form $\mathrm{diag}(1, \ldots, 1, 0, \ldots, 0)$.

6. Show that, with a rotation (in the xy-plane) of angle θ written as
$$I_\theta = \begin{pmatrix} \cos\theta & \sin\theta \\ -\sin\theta & \cos\theta \end{pmatrix},$$

(a) $(I_\theta)^T = I_{-\theta}$,

(b) $I_\theta I_\phi = I_\phi I_\theta = I_{\theta+\phi}$,

(c) $I_\theta^n = I_{n\theta}$,

(d) a reflection is expressed as $\begin{pmatrix} 1 & 0 \\ 0 & -1 \end{pmatrix} I_\theta$.

7. If A is a real orthogonal matrix with $\det A = -1$, show that A has an eigenvalue -1.

8. Let A be a 3×3 real orthogonal matrix with $\det A = 1$. Show that
$$(\operatorname{tr} A - 1)^2 + \sum_{i<j}(a_{ij} - a_{ji})^2 = 4.$$

9. Let A be an $n \times n$ real matrix. Denote $\Sigma(A) = \sum_{i,j=1}^n a_{ij}^2$. Show that A is real orthogonal if and only if
$$\Sigma(A^T X A) = \Sigma(X), \quad \text{for all } n \times n \text{ real } X.$$

10. If $A \in \mathbb{M}_n$ is real symmetric and idempotent, show that $0 \leq a_{ii} \leq 1$ for each i and $|a_{ij}| \leq \frac{1}{2}$ for all $i \neq j$. Moreover, if $a_{ii} = 0$ or 1, then $a_{ij} = a_{ji} = 0$ for all j with $j \neq i$.

11. Let A be an $n \times n$ real orthogonal matrix such that $\operatorname{rank}(A - I_n) = 1$. Show that A is real orthogonally similar to $\operatorname{diag}(-1, 1, \ldots, 1)$.

12. Let $A = (a_{ij})$ be an $n \times n$ real matrix and C_{ij} be the cofactor of a_{ij}, $i, j = 1, 2, \ldots, n$. Show that A is orthogonal if and only if $\det A = \pm 1$ and $a_{ij} = C_{ij}$ if $\det A = 1$, $a_{ij} = -C_{ij}$ if $\det A = -1$ for all i, j.

13. Let $A = (a_{ij}) \neq 0$ be an $n \times n$ real matrix and C_{ij} be the cofactor of a_{ij}, $i, j = 1, 2, \ldots, n$. If $n > 2$, show that A is orthogonal if $a_{ij} = C_{ij}$ for all i, j, or $a_{ij} = -C_{ij}$ for all i, j.

14. Let
$$A = \frac{1}{\sqrt{2}}\begin{pmatrix} 0 & 0 & -1 & 1 \\ 0 & 0 & 1 & 1 \\ 1 & 1 & 0 & 0 \\ 1 & -1 & 0 & 0 \end{pmatrix}, \quad U = \begin{pmatrix} 0 & 1 & 0 & 0 \\ -1 & 0 & 0 & 0 \\ 0 & 0 & 0 & 1 \\ 0 & 0 & -1 & 0 \end{pmatrix}.$$

(a) Show that $UA = AU^T$.

(b) Find the eigenvalues of U.

(c) Show that $A^2 \neq I$.

5.3 Metric Space and Contractions

A *metric space* consists of a set M and a mapping

$$d : M \times M \mapsto \mathbb{R},$$

called a *metric* of M, for which

1. $d(x, y) \geq 0$, and $d(x, y) = 0$ if and only if $x = y$,

2. $d(x, y) = d(y, x)$ for all x and y in M, and

3. $d(x, y) \leq d(x, z) + d(z, y)$ for all x, y, and z in M.

Consider a sequence of points $\{x_i\}$ in a metric space M. If for every $\epsilon > 0$ there exists a positive integer N such that $d(x_i, x_j) < \epsilon$ for all $i, j > N$, then the sequence is called a *Cauchy sequence*. A sequence $\{x_i\}$ *converges* to a point x if for every $\epsilon > 0$ there exists a positive integer N such that $d(x, x_i) < \epsilon$ for all $i > N$. A metric space M is said to be *complete* if every Cauchy sequence converges to a point of M. For instance, $\{c^n\}$, $0 < c < 1$, is a Cauchy sequence of the complete metric space $\mathbb{R}$ with metric $d(x, y) = |x - y|$.

$\mathbb{C}^n$ is a metric space with metric

$$d(x, y) = \|x - y\|, \quad x, y \in \mathbb{C}^n, \tag{5.3}$$

defined by the vector norm

$$\|x\| = \left(\sum_{i=1}^{n} |x_i|^2 \right)^{\frac{1}{2}}, \quad x \in \mathbb{C}^n.$$

Let $f : M \mapsto M$ be a mapping of a metric space M with metric d into itself. We call f a *contraction* if there exists a constant c with $0 < c \leq 1$ such that

$$d(f(x), f(y)) \leq c d(x, y), \quad \text{for all } x, y \in M. \tag{5.4}$$

If $0 < c < 1$, we say that f is a *strict contraction*.

SEC. 5.3 METRIC SPACE AND CONTRACTIONS 143

For the metric space $\mathbb{R}$ with the usual metric
$$d(x,y) = |x-y|, \quad x, y \in \mathbb{R},$$
the mapping $x \mapsto \sin\frac{x}{2}$ is a strict contraction, since by the sum-to-product trigonometric identity (Or use the mean value theorem.)
$$\sin\frac{x}{2} - \sin\frac{y}{2} = 2\cos\frac{x+y}{4}\sin\frac{x-y}{4},$$
together with the inequality $|\sin x| \leq |x|$, we have for all x, y in $\mathbb{R}$
$$\left|\sin\frac{x}{2} - \sin\frac{y}{2}\right| \leq \frac{1}{2}|x-y|.$$
The mapping $x \mapsto \sin x$ is a contraction, but not strict, for
$$\lim_{x \to 0} \frac{\sin x}{x} = 1.$$

A point x in a metric space M is referred to as a *fixed point* of a mapping f if $f(x) = x$. The following fixed-point theorem of a contraction has applications in many fields. For example, it gives a useful method for constructing solutions of differential equations.

Theorem 5.6 *Let $f : M \mapsto M$ be a strict contraction mapping of a complete metric space M into itself. Then f has one and only one fixed point. Moreover, for any point $x \in M$, the sequence*
$$x, \; f(x), \; f^2(x), \; f^3(x), \; \ldots$$
converges to the fixed point.

PROOF. Let x be a point in M. Denote $d(x, f(x)) = \delta$. By (5.4)
$$d(f^n(x), f^{n+1}(x)) \leq c^n \delta, \quad n \geq 1. \tag{5.5}$$
The series $\sum_{n=1}^{\infty} c^n$ converges to $\frac{c}{1-c}$ for every fixed c, $0 < c < 1$. Hence, the sequence $f^n(x), n = 1, 2, \ldots$, is a Cauchy sequence, since
$$\begin{aligned}d(f^m(x), f^n(x)) &\leq d(f^m(x), f^{m+1}(x)) + \cdots + d(f^{n-1}(x), f^n(x)) \\ &\leq (c^m + \cdots + c^{n-1})\delta.\end{aligned}$$

Thus, the limit $\lim_{n\to\infty} f^n(x)$ exists in M, for M is complete. Let the limit be X. We show that X is the fixed point. Note that a contraction mapping is a continuous function by (5.4). Therefore,
$$f(X) = f(\lim_{n\to\infty} f^n(x)) = \lim_{n\to\infty} f^{n+1}(x) = X.$$
If $Y \in M$ is also a fixed point of f, then
$$d(X,Y) = d(f(X), f(Y)) \leq cd(X,Y).$$
It follows that $d(X,Y) = 0$, that is, $X = Y$ if $0 < c < 1$. ∎

Let A be an $m \times n$ complex matrix, and consider A as a mapping from $\mathbb{C}^n$ into itself defined by the ordinary matrix-vector product, namely, Ax, where $x \in \mathbb{C}^n$. Then inequality (5.4) is rewritten as
$$\|Ax - Ay\| \leq c\|x - y\|.$$
We show that A is a contraction if and only if $\sigma_{\max}(A)$, the largest singular value of A, does not exceed 1.

Theorem 5.7 *Matrix A is a contraction if and only if $\sigma_{\max}(A) \leq 1$.*

PROOF. Let A be $m \times n$. For any $x, y \in \mathbb{C}^n$, we have (Section 3.6)
$$\|Ax - Ay\| = \|A(x-y)\| \leq \sigma_{\max}(A)\|x - y\|.$$
It follows that A is a contraction if $\sigma_{\max}(A) \leq 1$. Conversely, suppose that A is a contraction; then for some c, $0 < c \leq 1$, and all $x, y \in \mathbb{C}^n$,
$$\|Ax - Ay\| \leq c\|x - y\|.$$
In particular, $\|Ax\| \leq c\|x\|$ for $x \in \mathbb{C}^n$. Thus, $\sigma_{\max}(A) \leq c \leq 1$. ∎

Note that unitary matrices are contractions, but not strict. One can also prove that a matrix A is a contraction if and only if
$$A^*A \leq I, \quad AA^* \leq I, \quad \text{or} \quad \begin{pmatrix} I & A \\ A^* & I \end{pmatrix} \geq 0.$$
Here $X \geq Y$, or $Y \leq X$, means $X - Y$ is positive semidefinite.

We conclude this section by presenting a result on partitioned positive semidefinite matrices, from which a variety of matrix inequalities can be derived.

Let A be a positive semidefinite matrix. Recall that $A^{\frac{1}{2}}$ is the square root of A, that is, $A^{\frac{1}{2}} \geq 0$ and $(A^{\frac{1}{2}})^2 = A$ (Section 3.2).

Theorem 5.8 *Let L and M be positive semidefinite matrices. Then*

$$\begin{pmatrix} L & X \\ X^* & M \end{pmatrix} \geq 0 \quad \Leftrightarrow \quad X = L^{\frac{1}{2}} C M^{\frac{1}{2}} \text{ for some contraction } C.$$

PROOF. Sufficiency: If $X = L^{\frac{1}{2}} C M^{\frac{1}{2}}$, then we write

$$\begin{pmatrix} L & X \\ X^* & M \end{pmatrix} = \begin{pmatrix} L^{\frac{1}{2}} & 0 \\ 0 & M^{\frac{1}{2}} \end{pmatrix} \begin{pmatrix} I & C \\ C^* & I \end{pmatrix} \begin{pmatrix} L^{\frac{1}{2}} & 0 \\ 0 & M^{\frac{1}{2}} \end{pmatrix}.$$

For the positive semidefiniteness, it suffices to note that (Problem 8)

$$\sigma_{\max}(C) \leq 1 \quad \Rightarrow \quad \begin{pmatrix} I & C \\ C^* & I \end{pmatrix} \geq 0.$$

For the other direction, assume that L and M are nonsingular and let $C = L^{-\frac{1}{2}} X M^{-\frac{1}{2}}$. Here the power $-\frac{1}{2}$ means the inverse of the square root. Then $X = L^{\frac{1}{2}} C M^{\frac{1}{2}}$ has the desired form. We need to show that C is a contraction. First notice that

$$C^* C = M^{-\frac{1}{2}} X^* L^{-1} X M^{-\frac{1}{2}}.$$

Notice also that, since the partitioned matrix is positive semidefinite,

$$P^* \begin{pmatrix} L & X \\ X^* & M \end{pmatrix} P = \begin{pmatrix} L & 0 \\ 0 & M - X^* L^{-1} X \end{pmatrix} \geq 0,$$

where

$$P = \begin{pmatrix} I & -L^{-1} X \\ 0 & I \end{pmatrix}.$$

Thus, $M - X^* L^{-1} X \geq 0$ (Problem 8, Section 3.2.). Therefore,

$$M^{-\frac{1}{2}}(M - X^* L^{-1} X) M^{-\frac{1}{2}} = I - M^{-\frac{1}{2}} X^* L^{-1} X M^{-\frac{1}{2}} \geq 0.$$

That is, $I - C^* C \geq 0$, and thus C is a contraction. The singular case of L and M follows from a continuity argument (Problem 16). ∎

We end this section by noting that targeting the submatrices X and X^* in the upper-right and lower-left corners in the given partitioned matrix by row and column elementary operations for block matrices is a basic technique in matrix theory. It will be used repeatedly in later chapters of this book.

Problems

1. Following the proof of Theorem 5.6, show that
$$d(x, X) \leq \frac{\delta}{1-c}.$$

2. Show that a contraction is a continuous function.

3. If f is a strict contraction of a complete metric space, show that
$$d(f^n(x), f^n(y)) \to 0, \quad \text{as } n \to \infty,$$
for any fixed x and y in the space.

4. Show that the product of contractions is again a contraction.

5. Is the mapping $x \mapsto \sin(2x)$ a contraction on $\mathbb{R}$? How about the mappings $x \mapsto 2\sin x$, $x \mapsto \frac{1}{2}\sin x$, and $x \mapsto 2\sin \frac{x}{2}$?

6. Construct an example of a map f for a metric space such that $d(f(x), f(y)) < d(x, y)$ for all $x \neq y$, but f has no fixed point.

7. If $A \in \mathbb{M}_n$ is a contraction with eigenvalues $\lambda(A)$, show that
$$|\det A| \leq 1, \quad |\lambda(A)| \leq 1, \quad |x^*Ax| \leq 1 \text{ for unit } x \in \mathbb{C}^n.$$

8. Show that an $m \times n$ matrix A is a contraction if and only if
   (a) $A^*A \leq I_n$,
   (b) $AA^* \leq I_m$,
   (c) $\begin{pmatrix} I & A \\ A^* & I \end{pmatrix} \geq 0$,
   (d) $x^*(A^*A)x \leq 1$ for every unit $x \in \mathbb{C}^n$, or
   (e) $\|Ax\| \leq \|x\|$ for every $x \in \mathbb{C}^n$.

9. Let A be an $n \times n$ matrix and B be an $m \times n$ matrix. Show that
$$\begin{pmatrix} A & B^* \\ B & I \end{pmatrix} \geq 0 \quad \Leftrightarrow \quad A \geq B^*B.$$

10. Let $A \in \mathbb{M}_n$. If $\sigma_{\max}(A) < 1$, show that $I_n - A$ is invertible.

11. Let A and B be n-square complex matrices. Show that
$$A^*A \leq B^*B$$
if and only if $A = CB$ for some contraction matrix C.

12. Consider the complete metric space $\mathbb{R}^2$ and let
$$A = \begin{pmatrix} \lambda & 0 \\ 0 & \lambda \end{pmatrix}, \quad \lambda \in (0,1).$$
Discuss the effect of an application of A to a nonzero vector $v \in \mathbb{R}^2$. Describe the geometric orbit of the iterates $A^n v$. What is the fixed point of A? What if the second λ in A is replaced with $\mu \in (0,1)$?

13. Let $A \in \mathbb{M}_n$ be a projection matrix, that is, $A^2 = A$. Show that $\|Ax\| \leq \|x\|$ for all $x \in \mathbb{C}^n$ if and only if A is Hermitian.

14. Let A and B be n-square positive definite matrices. Find the conditions on the invertible n-square matrix X so that
$$\begin{pmatrix} A & X^* \\ X & B \end{pmatrix} \geq 0 \quad \text{and} \quad \begin{pmatrix} A^{-1} & X^{-1} \\ (X^*)^{-1} & B^{-1} \end{pmatrix} \geq 0.$$

15. Let A be a matrix. If there exists a Hermitian matrix X such that
$$\begin{pmatrix} I+X & A \\ A^* & I-X \end{pmatrix} \geq 0,$$
show that
$$|(Ay,y)| \leq (y,y), \quad \text{for all } y.$$

16. Prove Theorem 5.8 for the singular case.

17. Show that for any matrices X and Y of the same size,
$$X + Y = (I + XX^*)^{\frac{1}{2}} C (I + Y^*Y)^{\frac{1}{2}}$$
for some contraction C. Derive the matrix inequality
$$|\det(X+Y)|^2 \leq \det(I + XX^*) \det(I + Y^*Y).$$

· ———— ⊙ ———— ·

5.4 Contractions and Unitary Matrices

The goal of this section is to present two theorems connecting contractions and unitary matrices. We focus on square matrices, for otherwise we can augment by zero entries to make the matrices square. We shall show that a matrix is a contraction if and only if it can be embedded in a unitary matrix, and if and only if it is a (finite) convex combination of unitary matrices.

Theorem 5.9 *A matrix A is a contraction if and only if*

$$U = \begin{pmatrix} A & X \\ Y & Z \end{pmatrix}$$

is unitary for some matrices X, Y, and Z of appropriate sizes.

PROOF. The sufficiency is easy to see, since if U is unitary, then

$$U^*U = I \;\Rightarrow\; A^*A + Y^*Y = I \;\Rightarrow\; A^*A \leq I.$$

Thus, A is a contraction. For the necessity, we take

$$U = \begin{pmatrix} A & (I - AA^*)^{\frac{1}{2}} \\ (I - A^*A)^{\frac{1}{2}} & -A^* \end{pmatrix}$$

and show that U is a unitary matrix as follows.

Let $A = VDW$ be a singular value decomposition of A, where V and W are unitary, and D is a nonnegative diagonal matrix with diagonal entries (singular values of A) not exceeding 1. Then

$$(I - AA^*)^{\frac{1}{2}} = V(I - D^2)^{\frac{1}{2}}V^*, \quad (I - A^*A)^{\frac{1}{2}} = W^*(I - D^2)^{\frac{1}{2}}W.$$

Since D is diagonal, it is easy to see that

$$D(I - D^2)^{\frac{1}{2}} = (I - D^2)^{\frac{1}{2}}D.$$

Multiplying by V and W from the left and right gives

$$VD(I - D^2)^{\frac{1}{2}}W = V(I - D^2)^{\frac{1}{2}}DW$$

SEC. 5.4   CONTRACTIONS AND UNITARY MATRICES   149

or equivalently
$$A(I - A^*A)^{\frac{1}{2}} = (I - AA^*)^{\frac{1}{2}}A.$$
With this result, a simple computation results in $U^*U = I$. ∎

Recall from Problem 9 of Section 3.6 that for any $A, B \in \mathbb{M}_n$
$$\sigma_{\max}(A + B) \le \sigma_{\max}(A) + \sigma_{\max}(B).$$
Thus, for unitary matrices U and V of the same size and $t \in (0, 1)$,
$$\sigma_{\max}(tU + (1-t)V) \le t\sigma_{\max}(U) + (1-t)\sigma_{\max}(V) = 1.$$
In other words, the matrix $tU + (1-t)V$, a convex combination of unitary matrices U and V, is a contraction.

Inductively, a convex combination of unitary matrices is a contraction (Problem 3). The converse is also true.

Theorem 5.10 *A matrix A is a contraction if and only if A is a finite convex combination of unitary matrices.*

PROOF. As discussed earlier, a convex combination of unitary matrices is a contraction. Let A be a contraction. We show that A is a convex combination of unitary matrices. The proof goes as follows: A is a convex combination of matrices $\mathrm{diag}(1, \ldots, 1, 0, \ldots, 0)$; each matrix in such a form is a convex combination of diagonal (unitary) matrices with diagonal entries ± 1. We then reach the conclusion that A is a convex combination of unitary matrices.

Let A be of rank r and $A = UDV$ be a singular value decomposition of A, where U and V are unitary, $D = \mathrm{diag}(\sigma_1, \ldots, \sigma_r, 0, \ldots, 0)$ with $1 \ge \sigma_1 \ge \sigma_2 \ge \cdots \ge \sigma_r > 0$.

If D is a convex combination of unitary matrices, say, W_i, then A is a convex combination of unitary matrices UW_iV. We may thus consider the diagonal matrix $A = \mathrm{diag}(\sigma_1, \ldots, \sigma_r, 0, \ldots, 0)$. Write

$$\begin{aligned}
A &= \mathrm{diag}(\sigma_1, \ldots, \sigma_r, 0, \ldots, 0) \\
&= (1 - \sigma_1)0 + (\sigma_1 - \sigma_2)\mathrm{diag}(1, 0, \ldots, 0) \\
&\quad + (\sigma_2 - \sigma_3)\mathrm{diag}(1, 1, 0, \ldots, 0) + \cdots \\
&\quad + (\sigma_{r-1} - \sigma_r)\mathrm{diag}(\overbrace{1, \ldots, 1}^{r-1}, 0, \ldots, 0) \\
&\quad + \sigma_r \mathrm{diag}(\overbrace{1, \ldots, 1}^{r}, 0, \ldots, 0).
\end{aligned}$$

That is, matrix A is a (finite) convex combination of matrices $E_i = \mathrm{diag}(1, \ldots, 1, 0, \ldots, 0)$ with i copies of 1, where $0 \le i \le r$. We now show that such a matrix is a convex combination of diagonal matrices with entries ± 1. Let

$$F_i = \mathrm{diag}(0, \ldots, 0, \overbrace{-1, \ldots, -1}^{n-i}).$$

Then

$$E_i = \frac{1}{2}I + \frac{1}{2}(E_i + F_i)$$

is a convex combination of unitary matrices I and $E_i + F_i$.

It follows that if $\sigma_1 \le 1$, then the matrix $\mathrm{diag}(\sigma_1, \ldots, \sigma_r, 0, \ldots, 0)$, thus A, is a convex combination of diagonal matrices in the form $\mathrm{diag}(1, \ldots, 1, 0, \ldots, 0)$, which in turn is a convex combination of (diagonal) unitary matrices. The proof is complete by Problem 11. ∎

Problems

1. Let $t \in [0, 1]$. Write t as a convex combination of 1 and -1. Write matrix A as a convex combination of unitary matrices, where

$$A = \begin{pmatrix} \frac{1}{2} & 0 \\ 0 & \frac{1}{3} \end{pmatrix}.$$

2. Let λ and μ be positive numbers. Show that for any $t \in [0,1]$

$$\lambda\mu \le \big(t\lambda + (1-t)\mu\big)\big(t\mu + (1-t)\lambda\big).$$

3. Show by induction that a finite convex combination of unitary matrices of the same size is a contraction.

4. For any two matrices U and V of the same size, show that

$$U^*V + V^*U \le U^*U + V^*V.$$

In particular, if U and V are unitary, then

$$U^*V + V^*U \le 2I.$$

Also show that for any $t \in [0, 1]$ and unitary U and V,

$$\big(tU + (1-t)V\big)^*\big(tU + (1-t)V\big) \le I.$$

5. Prove or disprove that a convex combination, the sum, or the product of two unitary matrices is a unitary matrix.

6. If A is a contraction satisfying $A + A^* = 2I$, show that $A = I$.

7. Let A be a complex contraction matrix. Show that

$$\begin{pmatrix} A & (I - AA^*)^{\frac{1}{2}} \\ -(I - A^*A)^{\frac{1}{2}} & A^* \end{pmatrix}$$

and

$$\begin{pmatrix} (I - AA^*)^{\frac{1}{2}} & A \\ -A^* & (I - A^*A)^{\frac{1}{2}} \end{pmatrix}$$

are unitary matrices.

8. Let B_m be the $m \times m$ backward identity matrix. Show that

$$\frac{1}{\sqrt{2}} \begin{pmatrix} I_m & I_m \\ B_m & -B_m \end{pmatrix} \quad \text{and} \quad \frac{1}{\sqrt{2}} \begin{pmatrix} I_m & 0 & I_m \\ 0 & \sqrt{2} & 0 \\ B_m & 0 & -B_m \end{pmatrix}$$

are $m \times m$ and $(m+1) \times (m+1)$ unitary matrices, respectively.

9. Show that $\text{diag}(\sigma_1, \ldots, \sigma_r, 0, \ldots, 0)$ is a convex combination of matrices $\text{diag}(\sigma_1, \ldots, \sigma_1, 0, \ldots, 0)$ with k copies of σ_1, $k = 1, 2, \ldots, r$.

10. Show that $\text{diag}(\sigma, \ldots, \sigma, 0, \ldots, 0)$ is a convex combination of the following diagonal unitary matrices:

$$I, \quad G_i = \text{diag}(\overbrace{-1, \ldots, -1}^{i}, 1, \ldots, 1), \quad -G_i, \quad -I.$$

(Hint: Consider the cases for $0 \leq \sigma < \frac{1}{2}$ and $\frac{1}{2} \leq \sigma \leq 1$.)

11. Let $P_1, \ldots, P_m$ be a set of matrices. If each P_i is a convex combination of matrices $Q_1, \ldots, Q_n$, show that a convex combination of $P_1, \ldots, P_m$ is also a convex combination of $Q_1, \ldots, Q_n$.

5.5 The Unitary Similarity of Real Matrices

We show in this section that if two real matrices are similar over the complex number field $\mathbb{C}$, then they are similar over the real $\mathbb{R}$. The statement also holds for unitary similarity. Precisely, if two real matrices are unitarily similar, then they are real orthogonally similar.

Theorem 5.11 *Let A and B be real square matrices of the same size. If P is a complex invertible matrix such that $P^{-1}AP = B$, then there exists a real invertible matrix Q such that $Q^{-1}AQ = B$.*

PROOF. Write $P = P_1 + P_2 i$, where P_1 and P_2 are real square matrices. If $P_2 = 0$, we have nothing to show. Otherwise, by rewriting $P^{-1}AP = B$ as $AP = PB$, we have $AP_1 = P_1 B$ and $AP_2 = P_2 B$. It follows that for any real number t

$$A(P_1 + tP_2) = (P_1 + tP_2)B.$$

Since $\det(P_1 + tP_2) = 0$ for a finite number of t, we can choose a real t so that the matrix $Q = P_1 + tP_2$ is invertible. Thus, A and B are similar via the real invertible matrix Q. ∎

For the unitary similarity, we begin with a result that is of interest in its own right.

Theorem 5.12 *Let U be a symmetric unitary matrix, that is, $U^T = U$ and $U^* = U^{-1}$. Then there exists a complex matrix S satisfying*

1. *$S^2 = U$,*

2. *S is unitary,*

3. *S is symmetric,*

4. *S commutes with every matrix that commutes with U.*

In other words, every symmetric unitary U has a symmetric unitary square root that commutes with any matrix commuting with U.

PROOF. Since U is unitary, it is unitarily diagonalizable (Theorem 5.1). Let $U = VDV^*$, where V is unitary and $D = a_1 I_1 \oplus \cdots \oplus a_k I_k$ with all a_j distinct and I_i identity matrices of certain sizes. Since U is unitary and hence has eigenvalues of modulus 1, we write each $a_j = e^{i\theta_j}$ for some θ_j real.

Now let $S = V(b_1 I_1 \oplus \cdots \oplus b_k I_k) V^*$, where $b_j = e^{i\theta_j/2}$. Obviously S is a unitary matrix and $S^2 = U$.

If A is a matrix commuting with U, then $V^* AV$ commutes with D. It follows that $V^* AV = A_1 \oplus \cdots \oplus A_k$, with each A_i having the same size as I_i (Problem 4). Thus, S commutes with A.

Since $U = U^T$, this implies that $V^T V$ commutes with D, so that $V^T V$ commutes with $b_1 I_1 \oplus \cdots \oplus b_k I_k$. Thus, S is symmetric. ∎

Theorem 5.13 *Let A and B be real square matrices of the same size. If $A = UBU^*$ for some unitary matrix U, then there exists a real orthogonal matrix Q such that $A = QBQ^T$.*

PROOF. Since A and B are real, we have $UBU^* = A = \overline{A} = \overline{U} B U^T$. This gives $U^T U B = B U^T U$. Now that $U^T U$ is symmetric unitary, by the preceding theorem it has a symmetric unitary square root, say S, that is, $U^T U = S^2$, which commutes with B.

Let $Q = US^{-1}$ or $U = QS$. Then Q is also unitary. Notice that

$$Q^T Q = (US^{-1})^T (US^{-1}) = S^{-1} U^T U S^{-1} = I.$$

Hence Q is orthogonal. Q is real, for $Q^T = Q^{-1} = Q^*$ yields $Q = \overline{Q}$. Putting it all together, S and B commute, S is unitary, and Q is real orthogonal. We thus have

$$\begin{aligned} A &= UBU^* = (US^{-1})(SB)U^* = Q(BS)U^* \\ &= QB(S^{-1})^* U^* = QBQ^* = QBQ^T. \end{aligned}$$ ∎

Problems

1. If A^2 is a unitary matrix, is A necessarily a unitary matrix?

2. If A is an invertible matrix with complex, real, rational, or integer entries, is the inverse of A also a matrix with complex, real, rational, or integer entries, respectively?

3. Let A be a normal matrix. Show that there exists a normal matrix B such that $B^2 = A$. Is such a B unique?

4. If matrix A commutes with $B = b_1 I_1 \oplus \cdots \oplus b_k I_k$, where the I_i are identity matrices and all b_i are distinct, show that A is of the form $A = A_1 \oplus \cdots \oplus A_k$, where each A_i has the same size as the corresponding I_i.

5. Show that A and B are similar via a real invertible matrix Q, where

$$A = \begin{pmatrix} 1 & 1 \\ 0 & 0 \end{pmatrix}, \quad B = \begin{pmatrix} 1 & 0 \\ 0 & 0 \end{pmatrix}.$$

Find Q. Are they real orthogonally (or unitarily) similar?

6. If two matrices A and B with rational entries are similar over the complex $\mathbb{C}$, are they similar over the real $\mathbb{R}$? The rational $\mathbb{Q}$?

7. Let Q be a real orthogonal matrix. If λ is a real eigenvalue of Q and $u = x + yi$ is a corresponding eigenvector, where x and y are real, show that x and y are orthogonal and have the same length.

8. Find the conditions on the complex numbers a, b, c, and d so that

$$A = \begin{pmatrix} a & b \\ c & d \end{pmatrix}$$

is normal or symmetric unitary.

9. Show that if A is a real symmetric matrix, then there exists a real orthogonal Q such that $Q^T A Q$ is real diagonal. Give an example of a real normal matrix that is unitarily similar to a (complex) diagonal matrix but is not real orthogonally similar to a diagonal matrix.

10. Let

$$A = \begin{pmatrix} 1 & 0 & 1 \\ 0 & 2 & 1 \\ 0 & 0 & 2 \end{pmatrix}, \quad B = \begin{pmatrix} 1 & 0 & 0 \\ 0 & 2 & \pm\sqrt{2} \\ 0 & 0 & 2 \end{pmatrix}.$$

(a) What are the eigenvalues and eigenvectors of A and B?

(b) Why are A and B similar?

(c) Show that 1 is a singular value of B but not of A.

(d) Show that A cannot be unitarily similar to a direct sum of upper-triangular matrices of order 1 or 2.

(e) Can A and B be unitarily similar?

5.6 A Trace Inequality of Unitary Matrices

The set of all complex matrices of the same size forms an inner product space over $\mathbb{C}$ with respect to the inner product defined as

$$(A, B)_{\mathbb{M}} = \operatorname{tr}(B^*A).$$

In what follows we consider the inner product space $\mathbb{M}_n$ over $\mathbb{C}$ and present a trace inequality for complex unitary matrices, relating the average of the eigenvalues of each of two unitary matrices to that of their product. For this purpose, we first show an inequality for an inner product space V, which is of interest in its own right.

Theorem 5.14 *Let u, v, and w be unit vectors in V. Then*

$$\sqrt{1 - |(u,v)|^2} \leq \sqrt{1 - |(u,w)|^2} + \sqrt{1 - |(w,v)|^2} \,. \tag{5.6}$$

Equality holds if and only if w is a multiple of u or v.

PROOF. To prove this, we first notice that any component of w that is orthogonal to the span of u and v plays no role in (5.6), namely, we really have a problem in which u and v are arbitrary unit vectors, w is in the span of u and v, and $(w, w) \leq 1$. The case $w = 0$ is trivial. If $w \neq 0$, scaling up w to have length 1 diminishes the right-hand side of (5.6), so we are done if we can prove inequality (5.6) for arbitrary unit vectors u, v, and w with w in the span of u and v. The case in which u and v are dependent is trivial. Suppose u and v are linearly independent, and let $\{u, z\}$ be an orthonormal basis of $\operatorname{Span}\{u, v\}$, so that $v = \mu u + \lambda z$ and $w = \alpha u + \beta z$ for some complex numbers μ, λ, α, and β. Then we have

$$|\lambda|^2 + |\mu|^2 = 1 \quad \text{and} \quad |\alpha|^2 + |\beta|^2 = 1.$$

Use these relations and the arithmetic-geometric mean inequality, together with $|c| \geq \operatorname{Re}(c)$ for any complex number c, to compute

$$
\begin{aligned}
|\lambda\beta| &= \frac{1}{2}|\lambda\beta|(|\mu|^2 + |\lambda|^2 + |\alpha|^2 + |\beta|^2) \\
&\geq |\lambda\beta|(|\lambda\beta| + |\alpha\mu|) \\
&= |\lambda\beta|^2 + |\lambda\beta\alpha\mu| \\
&= |\lambda\beta|^2 + |\lambda\bar{\beta}\alpha\bar{\mu}| \\
&\geq |\lambda\beta|^2 + \operatorname{Re}(\lambda\bar{\beta}\alpha\bar{\mu}),
\end{aligned}
$$

so that $-2|\lambda\beta| \leq -2|\lambda\beta|^2 - 2\operatorname{Re}(\lambda\bar{\beta}\alpha\bar{\mu})$. Thus, we have

$$
\begin{aligned}
(|\lambda| - |\beta|)^2 &= |\lambda|^2 - 2|\lambda\beta| + |\beta|^2 \\
&\leq |\lambda|^2 + |\beta|^2 - 2|\lambda\beta|^2 - 2\operatorname{Re}(\lambda\bar{\beta}\alpha\bar{\mu}) \\
&= |\lambda|^2 + |\beta|^2(1 - |\lambda|^2) - |\lambda\beta|^2 - 2\operatorname{Re}(\lambda\bar{\beta}\alpha\bar{\mu}) \\
&= (1 - |\mu|^2) + |\beta|^2|\mu|^2 - |\lambda\beta|^2 - 2\operatorname{Re}(\lambda\bar{\beta}\alpha\bar{\mu}) \\
&= 1 - |\mu|^2(1 - |\beta|^2) - |\lambda\beta|^2 - 2\operatorname{Re}(\lambda\bar{\beta}\alpha\bar{\mu}) \\
&= 1 - |\mu\alpha|^2 - |\lambda\beta|^2 - 2\operatorname{Re}(\lambda\bar{\beta}\alpha\bar{\mu}) \\
&= 1 - |\alpha\bar{\mu} + \beta\bar{\lambda}|^2.
\end{aligned}
$$

This gives
$$
|\lambda| - |\beta| \leq \sqrt{1 - |\alpha\bar{\mu} + \beta\bar{\lambda}|^2},
$$
or
$$
|\lambda| \leq |\beta| + \sqrt{1 - |\alpha\bar{\mu} + \beta\bar{\lambda}|^2},
$$
which is the same as
$$
\sqrt{1 - |\mu|^2} \leq \sqrt{1 - |\alpha|^2} + \sqrt{1 - |\alpha\bar{\mu} + \beta\bar{\lambda}|^2}.
$$

Since $|\mu|^2 = |(u,v)|^2$, $|\alpha|^2 = |(u,w)|^2$, and
$$
|\alpha\bar{\mu} + \beta\bar{\lambda}|^2 = |(\alpha u + \beta z, \mu u + \lambda z)|^2 = |(w,v)|^2,
$$
the inequality (5.6) is proved.

Equality holds for the overall inequality if and only if equality holds at the two points in our derivation where we invoked the arithmetic-geometric mean inequality and $|c| \geq \operatorname{Re}(c)$. Thus, equality holds if and only if $|\lambda| = |\beta|$ and $|\alpha| = |\mu|$, as well as $\operatorname{Re}(\lambda\bar{\beta}\alpha\bar{\mu}) =$

$|\lambda\bar{\beta}\alpha\bar{\mu}|$. The former is equivalent to having $\lambda = e^{i\theta}\beta$ and $\mu = e^{i\phi}\alpha$ for some real numbers θ and ϕ, while the latter is then equivalent to $\text{Re}(|\alpha\beta|^2(e^{i(\theta-\phi)} - 1)) = 0$. Thus, $\alpha = 0$, $\beta = 0$, or $e^{i\theta} = e^{i\phi}$, so equality in (5.6) holds if and only if either w is a multiple of u ($\beta = 0$) or w is a multiple of v ($\alpha = 0$ or $e^{i\theta} = e^{i\phi}$). ∎

Now consider the vector space $\mathbb{M}_n$ of all $n \times n$ complex matrices with the inner product $(A, B)_\mathbb{M} = \text{tr}(B^*A)$ for A and B in $\mathbb{M}_n$.

Let U and V be n-square unitary matrices. By putting

$$u = \frac{1}{\sqrt{n}}V, \quad v = \frac{1}{\sqrt{n}}U^*, \quad w = \frac{1}{\sqrt{n}}I$$

in (5.6), and writing $m(X) = \frac{1}{n}\text{tr}\,X$ for the average of the eigenvalues of the matrix $X \in \mathbb{M}_n$, we have the following result.

Theorem 5.15 *For any unitary matrices U and V,*

$$\sqrt{1 - |m(UV)|^2} \leq \sqrt{1 - |m(U)|^2} + \sqrt{1 - |m(V)|^2}$$

with equality if and only if U or V is a unitary scalar matrix.

Problems

1. Let U be an $m \times n$ matrix such that $U^*U = I_n$. Show that
$$\text{tr}(UAU^*) = \text{tr}\,A, \quad \text{for any } A \in \mathbb{M}_n.$$

2. Show that for any square matrix A and positive integers p and q
$$|\text{tr}\,A^{p+q}|^2 \leq \text{tr}\left((A^*)^p A^p\right) \text{tr}\left((A^*)^q A^q\right).$$

3. If U is an $n \times n$ nonscalar unitary matrix with eigenvalues $\lambda_1, \ldots, \lambda_n$, show that the following strict inequality holds:
$$\left|\frac{\lambda_1 + \cdots + \lambda_n}{n}\right| < 1.$$

4. Let V be any square submatrix of a unitary matrix U. Show that
$$|\lambda(V)| \leq 1,$$
where $\lambda(V)$ is any eigenvalue of V.

5. Let $D = \mathrm{diag}(a_1, \ldots, a_n)$. Show that for any n-square unitary U,
$$\min_i |a_i| \leq |\lambda(DU)| \leq \max_i |a_i|,$$
where $\lambda(DU)$ is any eigenvalue of DU.

6. With $\|A\| = \sqrt{(A,A)_{\mathrm{M}}} = \sqrt{\mathrm{tr}(A^*A)}$, show that for any A, $B \in \mathbb{M}_n$,
$$\|A + B\| \leq \|A\| + \|B\| \quad \text{and} \quad \|AB\| \leq \|A\|\,\|B\|.$$

7. With $\|A\| = \sqrt{(A,A)_{\mathrm{M}}} = \sqrt{\mathrm{tr}(A^*A)}$, show that for any square complex matrix A with (all) real eigenvalues,
$$\|A\|^2 = \left\|\frac{A+A^*}{2}\right\|^2 + \left\|\frac{A-A^*}{2}\right\|^2.$$

8. Let U be a unitary matrix. If λ and μ are two different eigenvalues of U, show that their eigenvectors u and v are orthogonal. Further show that $au + bv$ cannot be an eigenvector of U if $ab \neq 0$.

9. Let $\mathbb{P}^2$ be the collection of all the unit vectors in $\mathbb{C}^n$. Define
$$d(x,y) = \sqrt{1 - |(x,y)|^2}, \quad x, y \in \mathbb{P}^2.$$
Show that $d(x,y) = d(y,x)$ for all x and y in $\mathbb{P}^2$ and that $d(x,y) = 0$ if and only if $x = cy$ for some complex number c with $|c| = 1$.

10. For nonzero vectors u, $v \in \mathbb{C}^2$, define
$$d(u,v) = \sqrt{1 - \frac{|(u,v)|^2}{\|u\|^2 \|v\|^2}}.$$
Show that for any u, v, $w \in \mathbb{C}^2$, and λ, $\mu \in \mathbb{C}$,

(a) $d(\lambda u, \mu v) = d(u,v)$,

(b) $d(u,v) \leq d(u,w) + d(w,v)$,

(c) $d(u,v) = d(z_u, z_v)$, where $z_x = \frac{x_2}{x_1}$ if $x = (x_1, x_2)^T \in \mathbb{C}^2$ with $x_1 \neq 0$, and
$$d(z_u, z_v) = \frac{|z_u - z_v|}{\sqrt{(1 + |z_u|^2)(1 + |z_v|^2)}}.$$

CHAPTER 6

Positive Semidefinite Matrices

INTRODUCTION This chapter studies the positive semidefinite matrices, concentrating primarily on the inequalities of this type of matrix. The main goal is to present the fundamental results and show some often-used techniques. Section 6.1 gives the basic properties, Section 6.2 treats the Löwner partial ordering of positive semidefinite matrices, and Section 6.3 presents some inequalities of principal submatrices. Section 6.4 derives inequalities of partitioned positive semidefinite matrices using Schur complements, while Sections 6.5 and 6.6 investigate the Hadamard product and Kronecker product and related matrix inequalities. Finally, Section 6.7 shows matrix inequalities of the Cauchy-Schwarz type.

6.1 Positive Semidefinite Matrices

An n-square complex matrix A is said to be *positive semidefinite* or *nonnegative definite*, written as $A \geq 0$, if

$$x^*Ax \geq 0, \quad \text{for all } x \in \mathbb{C}^n. \tag{6.1}$$

A is further called *positive definite*, symbolized $A > 0$, if the strict inequality in (6.1) holds for all nonzero $x \in \mathbb{C}^n$.

It is immediate that if A is an $n \times n$ complex matrix, then

$$A \geq 0 \quad \Leftrightarrow \quad X^*AX \geq 0 \qquad (6.2)$$

for every $n \times m$ complex matrix X. (Note that one may augment a vector $x \in \mathbb{C}^n$ by zero entries to get a matrix of size $n \times m$.)

The following decomposition theorem (see the spectral decomposition theorem) of positive semidefinite matrices best characterizes positive semidefiniteness under unitary similarity.

Theorem 6.1 *An $n \times n$ complex matrix A is positive semidefinite if and only if there exists a unitary matrix U such that*

$$A = U^* \operatorname{diag}(\lambda_1, \ldots, \lambda_n) U,$$

where the λ_i are nonnegative. In addition,

$$A \geq 0 \Rightarrow \det A \geq 0 \quad \text{and} \quad A > 0 \Rightarrow \det A > 0.$$

Positive semidefinite matrices have many interesting and important properties and play a central role in matrix theory.

Theorem 6.2 *Let A be an n-square complex matrix. Then*

1. *A is positive definite if and only if the determinant of every leading principal submatrix (i.e., minor) of A is positive;*

2. *A is positive semidefinite if and only if the determinant of every (not only leading) principal submatrix of A is nonnegative.*

PROOF. Let A_k be a $k \times k$ principal submatrix of $A \in \mathbb{M}_n$. By permuting rows and columns we may place A_k in the upper-left corner of A. In other words, there exists a permutation matrix P such that A_k is the $(1, 1)$-block of $P^T A P$.

If $A \geq 0$, then (6.1) holds. Thus, for any $x \in \mathbb{C}^k$,

$$x^* A_k x = y^* A y \geq 0, \quad \text{where } y = P \begin{pmatrix} x \\ 0 \end{pmatrix} \in \mathbb{C}^n.$$

This says that A_k is positive semidefinite. Therefore, $\det A_k \geq 0$. The strict inequalities hold for positive definite matrix A.

Conversely, if every principal submatrix of A has a nonnegative determinant, then the polynomial in λ, by Problem 17 of Section 1.3,

$$\det(\lambda I - A) = \lambda^n - \delta_1 \lambda^{n-1} + \delta_2 \lambda^{n-2} - \cdots (-1)^n \det A,$$

has no negative zeros, since each δ_i, the sum of the determinants of all the principal matrices of order i, is nonnegative.

The case where A is positive definite follows similarly. ∎

Note that all the determinants of the leading principal submatrices of a matrix A being nonnegative does not imply that $A \geq 0$.

As a side product of the proof, we see that A is positive (semi)definite if and only if all of its principal submatrices are positive (semi)definite.

It is immediate that $A \geq 0 \Rightarrow a_{ii} \geq 0$ and that $a_{ii}a_{jj} \geq |a_{ij}|^2$ for $i \neq j$ by considering 2×2 principal submatrices

$$\begin{pmatrix} a_{ii} & a_{ij} \\ a_{ji} & a_{jj} \end{pmatrix} \geq 0.$$

Thus, if some diagonal entry $a_{ii} = 0$, then $a_{ij} = 0$ for all j, and thus $a_{hi} = 0$ for all h, since A is Hermitian. We conclude that some diagonal entry $a_{ii} = 0$ if and only if the row and the column containing a_{ii} consist entirely of 0.

Using Theorem 6.1 and the fact that any square matrix is a product of a unitary matrix and an upper-triangular matrix (QR factorization), one can prove (Problem 15) the next result.

Theorem 6.3 *The following statements for $A \in \mathbb{M}_n$ are equivalent:*

1. *A is positive semidefinite;*

2. *$A = B^*B$ for some matrix B;*

3. *$A = C^*C$ for some upper-triangular matrix C;*

4. *$A = D^*D$ for some upper-triangular matrix D with nonnegative diagonal entries;*

5. *$A = P^* \begin{pmatrix} I_r & 0 \\ 0 & 0 \end{pmatrix} P$ for some invertible matrix P, where r is the rank of A.*

Every nonnegative number has a unique nonnegative square root. The analog for matrices is the following.

Theorem 6.4 *For every $A \geq 0$, there exists a unique $B \geq 0$ so that*
$$B^2 = A.$$
Furthermore, B can be expressed as a polynomial in A.

PROOF. Let $A = U^* \operatorname{diag}(\lambda_1, \ldots, \lambda_n)U$, where U is unitary. Take
$$B = U^* \operatorname{diag}(\lambda_1^{\frac{1}{2}}, \ldots, \lambda_n^{\frac{1}{2}})U.$$
Then B is a positive semidefinite matrix and $B^2 = A$.

To show the uniqueness, suppose C is a positive semidefinite matrix satisfying $C^2 = A$. Since the eigenvalues of C are the nonnegative square roots of the eigenvalues of A, we may write
$$C = V \operatorname{diag}(\lambda_1^{\frac{1}{2}}, \ldots, \lambda_n^{\frac{1}{2}})V^*$$
for some unitary matrix V. Then the identity $C^2 = B^2 = A$ gives
$$T \operatorname{diag}(\lambda_1, \ldots, \lambda_n) = \operatorname{diag}(\lambda_1, \ldots, \lambda_n)T,$$
where $T = UV$. By computation, one derives for each pair i and j
$$t_{ij}\lambda_j = \lambda_i t_{ij},$$
which implies
$$t_{ij}\lambda_j^{\frac{1}{2}} = \lambda_i^{\frac{1}{2}} t_{ij}.$$
Hence,
$$T \operatorname{diag}(\lambda_1^{\frac{1}{2}}, \ldots, \lambda_n^{\frac{1}{2}}) = \operatorname{diag}(\lambda_1^{\frac{1}{2}}, \ldots, \lambda_n^{\frac{1}{2}})T.$$
By putting back UV for T, it follows that $B = C$.

To see that B is a polynomial of A, let $p(x)$ be a polynomial, by interpolation, such that $p(\lambda_i) = \sqrt{\lambda_i}$, $i = 1, \ldots, n$ (Problem 4, Section 4.4). Then it is easy to verify that $p(A) = B$. ∎

Such a matrix B is called the *square root* of A, denoted by $A^{\frac{1}{2}}$.

Note that A^*A is positive semidefinite for every complex matrix A and that the eigenvalues of $(A^*A)^{\frac{1}{2}}$ are the singular values of A. We shall further discuss the matrix $(A^*A)^{\frac{1}{2}}$ in Chapters 7 and 8.

Problems

1. Show that if A is a positive semidefinite matrix, then so are the matrices $\overline{A}$, A^T, $\mathrm{adj}(A)$, and A^{-1} if the inverse exists.

2. Let A be a positive semidefinite matrix. Show that $\mathrm{tr}\, A \geq 0$. Equality holds if and only if $A = 0$.

3. Let $A \in \mathbb{M}_n$ be positive semidefinite. Show that $(\det A)^{\frac{1}{n}} \leq \frac{1}{n} \mathrm{tr}\, A$.

4. Find a 2×2 nonsymmetric real matrix A such that $x^T A x \geq 0$ for every $x \in \mathbb{R}^2$. What if $x \in \mathbb{C}^2$?

5. For what real number t is the following $n \times n$ matrix with diagonal entries 1 and off-diagonal entries t positive semidefinite?
$$\begin{pmatrix} 1 & & t \\ & \ddots & \\ t & & 1 \end{pmatrix}.$$

6. For what $x, y, z \in \mathbb{C}$ are the following matrices positive semidefinite?
$$\begin{pmatrix} 1 & 1 & x \\ 1 & 1 & 1 \\ \bar{x} & 1 & 1 \end{pmatrix}, \quad \begin{pmatrix} 1 & 1 & -1 \\ 1 & 1 & y \\ -1 & \bar{y} & 1 \end{pmatrix}, \quad \begin{pmatrix} 1 & z & 0 \\ \bar{z} & 1 & 1 \\ 0 & 1 & 1 \end{pmatrix}.$$

7. Let $x \in \mathbb{C}, u, v \in \mathbb{C}^n, \alpha \in [0, 1]$. Show that (if the powers make sense)
$$\begin{pmatrix} |x|^{2\alpha} & x \\ \bar{x} & |x|^{2(1-\alpha)} \end{pmatrix} \geq 0, \quad \begin{pmatrix} u^*u & u^*v \\ v^*u & v^*v \end{pmatrix} \geq 0.$$

8. If $\lambda, \mu \in \mathbb{C}$, show that the following matrices are positive semidefinite:
$$\begin{pmatrix} |\lambda|^2 + 1 & \lambda + \mu \\ \bar{\lambda} + \bar{\mu} & |\mu|^2 + 1 \end{pmatrix}, \quad \begin{pmatrix} |\lambda|^2 & \lambda\mu \\ \bar{\lambda}\bar{\mu} & |\mu|^2 \end{pmatrix}, \quad \begin{pmatrix} |\lambda|^2 + |\mu|^2 & \lambda + \mu \\ \bar{\lambda} + \bar{\mu} & 2 \end{pmatrix}.$$

9. Show that if A is positive definite, so is a principal submatrix of A. Conclude that the diagonal entries of A are all positive.

10. Let A be positive definite. Show that the matrix with (i,j)-entry
$$\frac{a_{ij}}{\sqrt{a_{ii}a_{jj}}}, \quad i,j = 1,2,\ldots,n,$$
is positive definite. What are the diagonal entries of this matrix?

11. Let $A \geq 0$. Show that A can be written as a sum of rank 1 matrices
$$A = \sum_{i=1}^{k} u_i u_i^*,$$
where each u_i is a column vector and $k = \text{rank}(A)$.

12. Let $A \in \mathbb{M}_n$. If $x^* A x = 0$ for some $x \neq 0$, does it follow that $A = 0$? or $Ax = 0$? What if A is positive semidefinite?

13. Let A be an n-square positive semidefinite matrix. Show that
$$\lambda_{\min}(A) \leq x^* A x \leq \lambda_{\max}(A), \text{ for any unit } x \in \mathbb{C}^n.$$

14. Show that $A \geq 0$ if and only if $A = Q^* Q$ for some matrix Q with linearly independent rows. What is the size of Q?

15. Prove Theorem 6.3.

16. Let A be a Hermitian matrix. Show that all the eigenvalues of A lie in the interval $[a,b]$ if and only if $A - aI \geq 0$ and $bI - A \geq 0$ and that there exist $\alpha > 0$ and $\beta > 0$ such that $\alpha I + A > 0$ and $I + \beta A > 0$.

17. Find a matrix $A \in \mathbb{M}_n$ such that all of its principal submatrices of order not exceeding $n - 1$ are positive semidefinite, but A is not.

18. Find a Hermitian matrix A such that the leading minors are all non-negative, but A is not positive semidefinite.

19. Let $A \in \mathbb{M}_n$. Show that $A \geq 0$ if and only if every leading principal submatrix of A (including A) is positive semidefinite.

20. Let $A \in \mathbb{M}_n$ be a singular Hermitian matrix. If A contains a positive definite principal submatrix of order $n - 1$, show that $A \geq 0$.

21. Let A be an $n \times n$ positive definite matrix. Show that $\text{tr}\, A \, \text{tr}\, A^{-1} \geq n$.

22. Let A be a nonzero n-square matrix. If A is Hermitian and satisfies
$$\frac{\text{tr}\, A}{(\text{tr}\, A^2)^{\frac{1}{2}}} \geq \sqrt{n-1},$$
show that $A \geq 0$. Conversely, if $A \geq 0$, show that
$$\frac{\text{tr}\, A}{(\text{tr}\, A^2)^{\frac{1}{2}}} \geq 1.$$

23. Find the square roots for the positive semidefinite matrices
$$\begin{pmatrix} 1 & 1 \\ 1 & 1 \end{pmatrix}, \quad \begin{pmatrix} 1 & \frac{1}{2} \\ \frac{1}{2} & 1 \end{pmatrix}, \quad \begin{pmatrix} a & 1 \\ 1 & a^{-1} \end{pmatrix}, \quad a > 0.$$

24. Does every normal matrix have a normal square root? Is it unique? How about a general complex matrix?

25. Let A be a Hermitian matrix. Show that there exists a unique Hermitian matrix B such that $A = B^3$. Show further that if $AP = PA$ for some matrix P, then $BP = PB$.

26. Let A be a square complex matrix such that $A + A^* \geq 0$. Show that
$$\det \frac{A + A^*}{2} \leq |\det A|.$$

27. Show that if B commutes with $A \geq 0$, then B commutes with $A^{\frac{1}{2}}$. Thus any positive semidefinite matrix commutes with its square root.

28. Let $A > 0$. Show that $(A^{-1})^{\frac{1}{2}} = (A^{\frac{1}{2}})^{-1}$, denoted by $A^{-\frac{1}{2}}$.

29. Let $A \geq 0$ and let B be a principal submatrix of A. Show that B is singular (i.e., $\det B = 0$) if and only if the rows (columns) of A that contain B are linearly dependent.

30. Let A be a Hermitian matrix. Show that neither A nor $-A$ is positive semidefinite if and only if at least one of the following holds:

    (a) A has a minor of even order with negative sign;

    (b) A has two minors of odd order with opposite signs.

31. Let A and B be $n \times n$ complex matrices. Show that
$$A^*A = B^*B$$
if and only if $B = UA$ for some unitary U. When does
$$\frac{A^* + A}{2} = (A^*A)^{\frac{1}{2}}?$$

6.2 A Pair of Positive Semidefinite Matrices

Inequality is one of the main topics in modern matrix theory. In this section we present some inequalities involving two positive semidefinite matrices.

Let A and B be two Hermitian matrices of the same size. If $A-B$ is positive semidefinite, we write

$$A \geq B \quad \text{or} \quad B \leq A.$$

It is easy to see that $\geq$ is a partial ordering, referred to as *Löwner partial ordering*, on the set of Hermitian matrices, that is,

1. $A \geq A$ for every Hermitian matrix A,

2. if $A \geq B$ and $B \geq A$, then $A = B$, and

3. if $A \geq B$ and $B \geq C$, then $A \geq C$.

The statement $A \geq 0 \Leftrightarrow X^*AX \geq 0$ in (6.2) of the previous section immediately generalizes as follows:

$$A \geq B \quad \Leftrightarrow \quad X^*AX \geq X^*BX \tag{6.3}$$

for every complex matrix X of appropriate size.

Theorem 6.5 *Let $A \geq 0$ and $B \geq 0$ be of the same size. Then*

 1. $A + B \geq B$,

 2. $A^{\frac{1}{2}} B A^{\frac{1}{2}} \geq 0$,

 3. $\operatorname{tr}(AB) \leq \operatorname{tr} A \operatorname{tr} B$,

 4. *the eigenvalues of AB are all nonnegative, Furthermore, AB is positive semidefinite if and only if $AB = BA$.*

SEC. 6.2   A PAIR OF POSITIVE SEMIDEFINITE MATRICES   167

PROOF. (1) is obvious. (2) follows from the observation that
$$A^{\frac{1}{2}}BA^{\frac{1}{2}} = (A^{\frac{1}{2}})^* BA^{\frac{1}{2}} \geq 0.$$

To show (3), by unitary similarity, with $A = U^*DU$,
$$\mathrm{tr}(AB) = \mathrm{tr}(U^*DUB) = \mathrm{tr}(DUBU^*),$$

we may assume that $A = \mathrm{diag}(\lambda_1, \ldots, \lambda_n)$. Suppose that $b_{11}, \ldots, b_{nn}$ are the diagonal entries of B. Then

$$\begin{aligned}
\mathrm{tr}(AB) &= \lambda_1 b_{11} + \cdots + \lambda_n b_{nn} \\
&\leq (\lambda_1 + \cdots + \lambda_n)(b_{11} + \cdots + b_{nn}) \\
&= \mathrm{tr}\,A\,\mathrm{tr}\,B.
\end{aligned}$$

For (4), recall that XY and YX have the same eigenvalues if X and Y are both square matrices of the same size. Thus,
$$AB = A^{\frac{1}{2}}(A^{\frac{1}{2}}B)$$

has the same eigenvalues as $A^{\frac{1}{2}}BA^{\frac{1}{2}}$, which is positive semidefinite.

AB is not positive semidefinite in general, since it need not be Hermitian. If A and B commute, however, then AB is Hermitian, for
$$(AB)^* = B^*A^* = BA = AB,$$

and thus $AB \geq 0$. Conversely, if $AB \geq 0$, then it is Hermitian, and
$$AB = (AB)^* = B^*A^* = BA. \blacksquare$$

What follows is the main result of this section, which we will use to reduce many problems involving a pair of positive definite matrices to one involving two diagonal matrices.

Theorem 6.6 *Let A and B be n-square positive semidefinite matrices. Then there exists an invertible matrix P such that*

$$P^*AP \quad \text{and} \quad P^*BP$$

are both diagonal matrices. In addition, if A is nonsingular, then P can be chosen so that $P^*AP = I$ and P^*BP is diagonal.

PROOF. Let $\operatorname{rank}(A+B) = r$ and S be a nonsingular matrix so that
$$S^*(A+B)S = \begin{pmatrix} I_r & 0 \\ 0 & 0 \end{pmatrix}.$$
Conformally partition S^*BS as
$$S^*BS = \begin{pmatrix} B_{11} & B_{12} \\ B_{21} & B_{22} \end{pmatrix}.$$
Then $S^*(A+B)S \geq S^*BS$ by Theorem 6.5(1) and (6.3). This implies
$$B_{22} = 0, \quad B_{12} = 0, \quad B_{21} = 0.$$
Now for B_{11}, since $B_{11} \geq 0$, there exists an $r \times r$ unitary matrix T such that $T^*B_{11}T$ is diagonal. Put
$$P = S \begin{pmatrix} T & 0 \\ 0 & I_{n-r} \end{pmatrix}.$$
Then P^*BP and $P^*AP = P^*(A+B)P - P^*BP$ are both diagonal.

If A is invertible, we write $A = C^*C$ for some matrix C. Consider matrix $(C^{-1})^*BC^{-1}$. Since it is positive semidefinite, we have a unitary matrix U such that
$$(C^{-1})^*BC^{-1} = UDU^*,$$
where D is a diagonal matrix with nonnegative diagonal entries.

Let $P = C^{-1}U$. Then $P^*AP = I$ and $P^*BP = D$. ∎

Many results can be derived by reduction of positive semidefinite matrices A and B to diagonal matrices, or further to nonnegative numbers, to which some elementary inequalities may apply. The following two are immediate from the previous theorem by writing $A = P^*D_1P$ and $B = P^*D_2P$, where P is an invertible matrix, and D_1 and D_2 are diagonal matrices with nonnegative entries.

Theorem 6.7 *Let $A \geq 0$, $B \geq 0$ be of the same order (> 1). Then*
$$\det(A+B) \geq \det A + \det B \tag{6.4}$$
with equality if and only if $A+B$ is singular or $A = 0$ or $B = 0$, and
$$(A+B)^{-1} \leq \frac{1}{4}(A^{-1} + B^{-1}) \tag{6.5}$$
if A and B are nonsingular, with equality if and only if $A = B$.

Theorem 6.8 *If $A \geq B \geq 0$. Then*

1. $\operatorname{rank}(A) \geq \operatorname{rank}(B)$,

2. $\det A \geq \det B$,

3. $B^{-1} \geq A^{-1}$ *if A and B are nonsingular.*

Every positive semidefinite matrix has a positive semidefinite square root. The square root is a matrix monotone function for positive semidefinite matrices in the sense that the Löwner partial ordering is preserved when taking the square root.

Theorem 6.9 *Let A and B be positive semidefinite matrices. Then*
$$A \geq B \quad \Rightarrow \quad A^{\frac{1}{2}} \geq B^{\frac{1}{2}}.$$

PROOF 1. It may be assumed that A is positive definite by continuity (Problem 4). Let $C = A^{\frac{1}{2}}$, $D = B^{\frac{1}{2}}$, and $E = C - D$. We have to establish $E \geq 0$. For this purpose, it is sufficient to show that the eigenvalues of E are all nonnegative. On one hand, notice that
$$0 \leq C^2 - D^2 = C^2 - (C-E)^2 = CE + EC - E^2.$$
It follows that $CE + EC \geq 0$, for E is Hermitian and $E^2 \geq 0$.

On the other hand, let λ be an eigenvalue of E and let u be an eigenvector corresponding to λ. Then λ is real and by (6.1),
$$0 \leq u^*(CE + EC)u = 2\lambda(u^*Cu).$$
Since $C > 0$, we have $\lambda \geq 0$. Hence $E \geq 0$, namely, $C \geq D$.

PROOF 2. First notice that $A^{\frac{1}{2}} - B^{\frac{1}{2}}$ is Hermitian. We show that the eigenvalues are all nonnegative. Let
$$(A^{\frac{1}{2}} - B^{\frac{1}{2}})x = \lambda x, \quad x \neq 0.$$
Then
$$B^{\frac{1}{2}}x = A^{\frac{1}{2}}x - \lambda x.$$
By the Cauchy-Schwarz inequality, for all $x, y \in \mathbb{C}^n$,
$$|x^*y| \leq (x^*x)^{\frac{1}{2}} (y^*y)^{\frac{1}{2}}.$$

We thus have

$$\begin{aligned}
x^*Ax &= (x^*Ax)^{\frac{1}{2}}(x^*Ax)^{\frac{1}{2}} \\
&\geq (x^*Ax)^{\frac{1}{2}}(x^*Bx)^{\frac{1}{2}} \\
&\geq x^*A^{\frac{1}{2}}B^{\frac{1}{2}}x \\
&= x^*A^{\frac{1}{2}}(A^{\frac{1}{2}}x - \lambda x) \\
&= x^*Ax - \lambda x^*A^{\frac{1}{2}}x.
\end{aligned}$$

It follows that $\lambda x^*A^{\frac{1}{2}}x \geq 0$ for all x in $\mathbb{C}$, so $\lambda \geq 0$.

PROOF 3. If A is positive definite, then by using (6.3) and by multiplying both sides of $B \leq A$ by $A^{-\frac{1}{2}} = (A^{-\frac{1}{2}})^*$, we obtain

$$A^{-\frac{1}{2}}BA^{-\frac{1}{2}} \leq I,$$

rewritten as

$$(B^{\frac{1}{2}}A^{-\frac{1}{2}})^*(B^{\frac{1}{2}}A^{-\frac{1}{2}}) \leq I,$$

which gives

$$\sigma_{\max}(B^{\frac{1}{2}}A^{-\frac{1}{2}}) \leq 1,$$

where $\sigma_{\max}$ means the largest singular value. Thus, by Problem 8,

$$\lambda_{\max}(A^{-\frac{1}{4}}B^{\frac{1}{2}}A^{-\frac{1}{4}}) = \lambda_{\max}(B^{\frac{1}{2}}A^{-\frac{1}{2}}) \leq \sigma_{\max}(B^{\frac{1}{2}}A^{-\frac{1}{2}}) \leq 1,$$

where $A^{-\frac{1}{4}}$ is the square root of $A^{-\frac{1}{2}}$, and, by Problem 9,

$$0 \leq A^{-\frac{1}{4}}B^{\frac{1}{2}}A^{-\frac{1}{4}} \leq I.$$

Multiplying both sides by $A^{\frac{1}{4}}$, the square root of $A^{\frac{1}{2}}$, we see that

$$B^{\frac{1}{2}} \leq A^{\frac{1}{2}}.$$

The case for singular A follows from a continuity argument. ∎

Problems _____

1. Show that $A \geq B \Rightarrow \operatorname{tr} A \geq \operatorname{tr} B$. When does equality occur?
2. Give an example where $A \geq 0$ and $B \geq 0$ but AB is not Hermitian.

3. Show by example that $A \geq B \geq 0 \Rightarrow A^2 \geq B^2$ is not true in general. But $AB = BA$ and $A \geq B \geq 0$ imply $A^k \geq B^k$, $k = 1, 2, \ldots$.

4. Show Theorem 6.9 for the singular case, by recalling the proof of Theorem 6.4. That is, $\lim_{\epsilon \to 0^+} (A + \epsilon I)^{\frac{1}{2}} = A^{\frac{1}{2}}$ for $A \geq 0$.

5. If A is an $n \times n$ complex matrix such that $x^*Ax \geq x^*x$ for every $x \in \mathbb{C}^n$, show that A is nonsingular and that $A \geq I \geq A^{-1} > 0$.

6. Let $A, B \in \mathbb{M}_n$. Show that $A > B$ (i.e., $A - B > 0$) if and only if $X^*AX > X^*BX$ for every $n \times m$ matrix X with rank$(X) = m$.

7. Let A be a nonsingular Hermitian matrix. Show that $A \geq A^{-1}$ if and only if every eigenvalue of A is greater than or equal to 1.

8. Let $A \in \mathbb{M}_n$. If $A \geq 0$, show that $x^*Ax \leq \lambda_{\max}(A)$ for all unit x and that $|\lambda(A)| \leq \sigma_{\max}(A)$ for every eigenvalue $\lambda(A)$ of A.

9. Let $A = A^*$. Show that $0 \leq A \leq I \Leftrightarrow$ every eigenvalue $\lambda(A) \in [0, 1]$.

10. Let $A \geq 0$. Show that $\lambda_{\max}(A)I \geq A$ and $\lambda_{\max}(A) \geq \max\{a_{ii}\}$.

11. Let $A \geq 0$ and $B \geq 0$ be of the same size. As is known, the eigenvalues of positive semidefinite matrices are the same as the singular values, and the eigenvalues of AB are nonnegative. Are the eigenvalues of AB in this case necessarily equal to the singular values of AB?

12. Show that the eigenvalues of the product of three positive semidefinite matrices are not necessarily nonnegative by the example

$$A = \begin{pmatrix} 1 & 1 \\ 1 & 1 \end{pmatrix}, \quad B = \begin{pmatrix} 2 & 1 \\ 1 & 1 \end{pmatrix}, \quad C = \begin{pmatrix} 2 & i \\ -i & 1 \end{pmatrix}.$$

13. Let A and B be n-square positive semidefinite matrices.

    (a) Show that $A > B \geq 0 \Leftrightarrow \lambda_{\max}(A^{-1}B) < 1$.

    (b) Show that $A > B \geq 0 \Rightarrow \det A > \det B$.

    (c) Give an example that $A \geq B \geq 0$, $\det A = \det B$, but $A \neq B$.

14. Let $A > 0$ and $B = B^*$ (not necessarily positive semidefinite) be of the same size. Show that there exists a nonsingular matrix P such that $P^*AP = I$, P^*BP is diagonal, and the diagonal entries of P^*BP are the eigenvalues of $A^{-1}B$. Show by example that the assertion is not true in general when $A > 0$ is replaced with $A \geq 0$.

15. Let A, B be $m \times n$ matrices. Show that for any $m \times m$ matrix $X > 0$

$$\operatorname{tr}(A^*B) \leq \operatorname{tr}(A^*XA) \operatorname{tr}(B^*X^{-1}B).$$

16. Let $A = (a_{ij})$ be an $n \times n$ positive semidefinite matrix. Show that

    (a) $\sum_{i=1}^n a_{ii}^2 \leq \operatorname{tr} A^2 \leq \left(\sum_{i=1}^n a_{ii}\right)^2$,

    (b) $\left(\sum_{i=1}^n a_{ii}\right)^{\frac{1}{2}} \leq \operatorname{tr} A^{\frac{1}{2}} \leq \sum_{i=1}^n a_{ii}^{\frac{1}{2}}$,

    (c) $\left(\sum_{i=1}^n a_{ii}\right)^{-1} \leq \operatorname{tr} A^{-1}$ if A positive definite.

    Is it true that $\operatorname{tr} A^{-1} \leq \sum_{i=1}^n a_{ii}^{-1}$?

17. Let $A \geq 0$ and $B \geq 0$ be of the same size. Show that
$$\operatorname{tr}(A^{\frac{1}{2}} B^{\frac{1}{2}}) \leq (\operatorname{tr} A)^{\frac{1}{2}} (\operatorname{tr} B)^{\frac{1}{2}}$$
    and
$$\left(\operatorname{tr}(A+B)\right)^{\frac{1}{2}} \leq (\operatorname{tr} A)^{\frac{1}{2}} + (\operatorname{tr} B)^{\frac{1}{2}}.$$

18. Let $A > 0$ and $B > 0$ be of the same size. Show that
$$\operatorname{tr}\left((A^{-1} - B^{-1})(A - B)\right) \leq 0.$$

19. Let $A > 0$ and $B \geq C \geq 0$ be all of the same size. Show that
$$\operatorname{tr}\left((A+B)^{-1} B\right) \geq \operatorname{tr}\left((A+C)^{-1} C\right).$$

20. Prove or disprove that for matrices A and B of the same size,
$$A > 0,\ B > 0 \quad \Rightarrow \quad AB + BA > 0.$$

21. Show that for Hermitian matrices A and B of the same size,
$$A^2 + B^2 - AB - BA \geq 0.$$

22. Let A and B be n-square Hermitian matrices. Show that
$$A > 0,\ AB + BA \geq 0 \quad \Rightarrow \quad B \geq 0$$
    and
$$A > 0,\ AB + BA > 0 \quad \Rightarrow \quad B > 0.$$
    Show that $A > 0$ in fact can be replaced by the weaker condition $A \geq 0$ with positive diagonal entries. Show by example that the assertions do not hold in general if $A > 0$ is replaced by $A \geq 0$.

23. Let $A > 0$ and $B > 0$ be of the same size. Show that
$$(A+B)^{-1} \leq A^{-1} \quad \text{(or } B^{-1}\text{)},$$
or equivalently,
$$(A^{-1} + B^{-1})^{-1} \leq A \quad \text{(or } B\text{)}.$$

24. Let $A > 0$ and $B > 0$ be of the same size. Show that
$$A^{-1} - (A+B)^{-1} - (A+B)^{-1} B (A+B)^{-1} \geq 0.$$

25. Let $A \geq 0$ and $B \geq 0$ be of the same size. Prove or disprove
$$\frac{1}{2}(A^2 + B^2) \geq \frac{1}{2}(AB + BA) \geq A^{\frac{1}{2}} B A^{\frac{1}{2}}.$$

26. Let $A \geq 0$ and $B \geq 0$ be of the same size. Prove or disprove that
$$(BAB)^\alpha = B^\alpha A^\alpha B^\alpha,$$
where $\alpha = 2$, $\frac{1}{2}$, or -1 if A and B are nonsingular.

27. Let $A \geq 0$ and $B \geq 0$ be of the same size. Show that
$$BA^2 B \leq I \quad \Rightarrow \quad B^{\frac{1}{2}} A B^{\frac{1}{2}} \leq I.$$

28. Let $A \geq 0$ and $B \geq 0$ be of order n, where $n > 1$. Show that
$$\left(\det(A+B)\right)^{\frac{1}{n}} \geq (\det A)^{\frac{1}{n}} + (\det B)^{\frac{1}{n}}.$$
Equality occurs if and only if $B = aA$ for some $a > 0$. Conclude that
$$\det(A+B) \geq \det A + \det B$$
with equality if and only if $A+B$ is singular or $A = 0$ or $B = 0$, and
$$\det(A+B) \geq \det A$$
with equality if and only if $A+B$ is singular or $B = 0$.

29. Let $A \geq 0$ and $B \geq 0$ be of the same size. Show that for any $\lambda, \mu \in \mathbb{C}$
$$|\det(\lambda A + \mu B)| \leq \det(|\lambda| A + |\mu| B).$$

30. Show by example that $A \geq B \geq 0$ does not imply
$$A^{\frac{1}{2}} - B^{\frac{1}{2}} \leq (A - B)^{\frac{1}{2}}.$$

31. Let $A \geq 0$ and $B \geq 0$ be of the same size. Show that
    (a) $\frac{A^{-1}+B^{-1}}{2} \geq \left(\frac{A+B}{2}\right)^{-1}$,
    (b) $\left(\frac{A+B}{2}\right)^{\frac{1}{2}} \geq \frac{A^{\frac{1}{2}}+B^{\frac{1}{2}}}{2}$,
    (c) $\left(\frac{A+B}{2}\right)^2 \leq \frac{A^2+B^2}{2}$,
    (d) $\left(\frac{A+B}{2}\right)^3 \leq \frac{A^3+B^3}{2}$, however, is not true in general.

32. Let $A \geq 0$ and $B \geq 0$ be of the same size. Show that for any $t \in [0,1]$ and $\tilde{t} = 1-t$, assuming that the involved inverses exist,
    (a) $(tA + \tilde{t}B)^{-1} \leq tA^{-1} + \tilde{t}B^{-1}$,
    (b) $(tA + \tilde{t}B)^{-\frac{1}{2}} \leq tA^{-\frac{1}{2}} + \tilde{t}B^{-\frac{1}{2}}$,
    (c) $(tA + \tilde{t}B)^{\frac{1}{2}} \geq tA^{\frac{1}{2}} + \tilde{t}B^{\frac{1}{2}}$,
    (d) $(tA + \tilde{t}B)^2 \leq tA^2 + \tilde{t}B^2$,
    (e) $\det(tA + \tilde{t}B) \geq (\det A)^t (\det B)^{\tilde{t}}$.

    Give the explicit formulas for $t = \frac{1}{2}$.

33. Let $A > 0$ and $B > 0$ be matrices of the same size with eigenvalues contained in the closed interval $[m, M]$. Show that
    $$\frac{2mM}{(m+M)^2}(A+B) \leq 2(A^{-1}+B^{-1})^{-1}$$
    $$\leq A^{\frac{1}{2}}(A^{-\frac{1}{2}}BA^{-\frac{1}{2}})^{\frac{1}{2}}A^{\frac{1}{2}}$$
    $$\leq \frac{1}{2}(A+B).$$

34. Show that $A \geq 0$ and $B \geq C \geq 0$ imply the matrix inequality
    $$A^{\frac{1}{2}}BA^{\frac{1}{2}} \geq A^{\frac{1}{2}}CA^{\frac{1}{2}}$$
    but not
    $$B^{\frac{1}{2}}AB^{\frac{1}{2}} \geq C^{\frac{1}{2}}AC^{\frac{1}{2}}.$$

35. Let A and B be n-square complex matrices. Prove or disprove
    $$A^*A \leq B^*B \quad \Rightarrow \quad A^*CA \leq B^*CB, \text{ for } C \geq 0.$$

36. Let A and B be n-square complex matrices. Prove or disprove
    $$A^*A \leq B^*B \quad \Rightarrow \quad AA^* \leq BB^*.$$

37. Let A, B, and C be Hermitian matrices of the same size. If $A \geq B$ and if C commutes with both AB and $A+B$, show that C commutes with A and B. What if the condition $A \geq B$ is removed?

6.3 Partitioned Positive Semidefinite Matrices

In this section we present the Fischer and Hadamard determinantal inequalities and the matrix inequalities involving principal submatrices of positive semidefinite matrices.

Let A be a square complex matrix partitioned as

$$A = \begin{pmatrix} A_{11} & A_{12} \\ A_{21} & A_{22} \end{pmatrix}, \qquad (6.6)$$

where A_{11} is a square submatrix of A. If A_{11} is nonsingular, we have

$$\begin{pmatrix} I & 0 \\ -A_{21}A_{11}^{-1} & I \end{pmatrix} A \begin{pmatrix} I & -A_{11}^{-1}A_{12} \\ 0 & I \end{pmatrix} = \begin{pmatrix} A_{11} & 0 \\ 0 & \widetilde{A_{11}} \end{pmatrix}, \qquad (6.7)$$

where

$$\widetilde{A_{11}} = A_{22} - A_{21}A_{11}^{-1}A_{12}$$

is called the *Schur complement* of A_{11} in A. By taking determinants,

$$\det A = \det A_{11} \, \det \widetilde{A_{11}}.$$

If A is a positive definite matrix, then A_{11} is nonsingular and

$$A_{22} \geq \widetilde{A_{11}} \geq 0.$$

The Fischer determinantal inequality follows, for $\det A_{22} \geq \det \widetilde{A_{11}}$.

Theorem 6.10 (Fischer Inequality) *If A is positive semidefinite matrix partitioned as in (6.6), then*

$$\det A \leq \det A_{11} \det A_{22}$$

with equality if and only if both sides vanish or $A_{12} = 0$. Also

$$|\det A_{12}|^2 \leq \det A_{11} \det A_{22}$$

if the blocks A_{11}, A_{12}, A_{21}, and A_{22} are square matrices of the same size.

PROOF. The nonsingular case follows from the earlier discussion. For the singular case, one may replace A with $A + \epsilon I$, $\epsilon > 0$, to obtain the desired inequality by a continuity argument.

If equality holds and both A_{11} and A_{22} are nonsingular, then

$$\det \widetilde{A_{11}} = \det A_{22} = \det(A_{22} - A_{21}A_{11}^{-1}A_{12}).$$

Thus, $A_{21}A_{11}^{-1}A_{12} = 0$ or $A_{12} = 0$ by Theorem 6.7 and Problem 9.

For the second inequality, notice that $A_{22} \geq A_{21}A_{11}^{-1}A_{12} \geq 0$. The assertion follows at once by taking the determinants. ∎

An induction on the size of the matrices gives the following result.

Theorem 6.11 (Hadamard Inequality) *Let A be a positive semidefinite matrix with diagonal entries a_{11}, a_{22}, ..., a_{nn}. Then*

$$\det A \leq a_{11}a_{22}\cdots a_{nn}.$$

Equality holds if and only if some $a_{ii} = 0$ or A is diagonal.

A direct proof goes as follows: Assume that each $a_{ii} > 0$ and let $D = \operatorname{diag}(a_{11}^{-\frac{1}{2}}, \ldots, a_{nn}^{-\frac{1}{2}})$. Put $B = DAD$. Then B is a positive semidefinite matrix with diagonal entries all equal to 1. By the arithmetic mean-geometric mean inequality, we have

$$n = \operatorname{tr} B = \sum_{i=1}^{n} \lambda_i(B) \geq n\left(\prod_{i=1}^{n} \lambda_i(B)\right)^{\frac{1}{n}} = n(\det B)^{\frac{1}{n}}.$$

This implies $\det B \leq 1$. Thus,

$$\det A = \det(D^{-1}BD^{-1}) = \prod_{i=1}^{n} a_{ii} \det B \leq \prod_{i=1}^{n} a_{ii}.$$

Equality occurs if and only if the eigenvalues of B are identical and $\det B = 1$, that is, B is the identity and A is diagonal.

It follows that for any complex matrix A of size $m \times n$,

$$\det(A^*A) \leq \prod_{j=1}^{n}\sum_{i=1}^{m} |a_{ij}|^2. \tag{6.8}$$

SEC. 6.3   PARTITIONED POSITIVE SEMIDEFINITE MATRICES   177

An interesting application of the Hadamard inequality is to show that if $A = B + Ci \geq 0$, where B and C are real matrices, then

$$\det A \leq \det B$$

by passing the diagonal entries of A to B through a real orthogonal diagonalization of the real matrix B (Problem 27).

We now turn our attention to the inequalities involving principal submatrices of positive semidefinite matrices.

Let A be an n-square positive semidefinite matrix. We denote in this section by $[A]_\omega$, or simply $[A]$, the $k \times k$ principal submatrix of A indexed by a sequence $\omega = \{i_1, \ldots, i_k\}$, where $1 \leq i_1 < \cdots < i_k \leq n$. We are interested in comparing $f([A])$ and $[f(A)]$, where $f(x)$ is the elementary function x^2, $x^{\frac{1}{2}}$, $x^{-\frac{1}{2}}$, or x^{-1}.

Theorem 6.12 *Let $A \geq 0$ and let $[A]$ be a principal submatrix of the matrix A. Then, assuming that the inverses involved exist,*

$$[A^2] \geq [A]^2, \quad [A^{\frac{1}{2}}] \leq [A]^{\frac{1}{2}}, \quad [A^{-\frac{1}{2}}] \geq [A]^{-\frac{1}{2}}, \quad [A^{-1}] \geq [A]^{-1}.$$

PROOF. We may assume that $[A] = A_{11}$ as in (6.6). Otherwise one can carry out a similar argument for $P^T A P$, where P is a permutation matrix so that $[A]$ is in the upper-left corner.

Partition A^2, $A^{\frac{1}{2}}$, and A^{-1} conformally to A in (6.6) as

$$A^2 = \begin{pmatrix} E & F \\ F^* & G \end{pmatrix}, \quad A^{\frac{1}{2}} = \begin{pmatrix} P & Q \\ Q^* & R \end{pmatrix}, \quad A^{-1} = \begin{pmatrix} X & Y \\ Y^* & Z \end{pmatrix}.$$

Upon computation of A^2 using (6.6), we have the first inequality:

$$[A^2] = E = A_{11}^2 + A_{12} A_{21} \geq A_{11}^2 = [A]^2.$$

This yields $[A]^{\frac{1}{2}} \geq [A^{\frac{1}{2}}]$ by replacing A with $A^{\frac{1}{2}}$ and then taking the square root. The third inequality, $[A^{-\frac{1}{2}}] \geq [A]^{-\frac{1}{2}}$, follows from an application of the last inequality to the second.

We have left to show $[A^{-1}] \geq [A]^{-1}$. By Theorem 2.4, we have

$$[A^{-1}] = X = A_{11}^{-1} + A_{11}^{-1} A_{12} \widetilde{A_{11}}^{-1} A_{21} A_{11}^{-1} \geq A_{11}^{-1} = [A]^{-1}. \quad \blacksquare$$

The inequalities in Theorem 6.12 may be unified in the form

$$[f(A)] \geq f([A]), \quad \text{where } f(x) = x^2, -x^{\frac{1}{2}}, x^{-\frac{1}{2}}, \text{ or } x^{-1}.$$

Notice that if A is an n-square positive definite matrix, then for any $n \times m$ matrix B,

$$\begin{pmatrix} A & B \\ B^* & B^*A^{-1}B \end{pmatrix} \geq 0. \tag{6.9}$$

Note also that $B^*A^{-1}B$ is the smallest matrix to make the block matrix positive semidefinite in the Löwner partial ordering sense.

Theorem 6.13 *Let $A \in \mathbb{M}_n$ be a positive definite matrix and let B be an $n \times m$ matrix. Then for any positive semidefinite $X \in \mathbb{M}_m$,*

$$\begin{pmatrix} A & B \\ B^* & X \end{pmatrix} \geq 0 \iff X \geq B^*A^{-1}B.$$

PROOF. It is sufficient to notice the matrix identity

$$\begin{pmatrix} I_n & 0 \\ -B^*A^{-1} & I_m \end{pmatrix} \begin{pmatrix} A & B \\ B^* & X \end{pmatrix} \begin{pmatrix} I_n & -A^{-1}B \\ 0 & I_m \end{pmatrix}$$

$$= \begin{pmatrix} A & 0 \\ 0 & X - B^*A^{-1}B \end{pmatrix}. \blacksquare$$

Note that, by Theorem 2.4, $[A^{-1}] = (A_{11} - A_{12}A_{22}A_{21})^{-1}$. Thus for any positive semidefinite matrix A partitioned as in (6.6),

$$A - \begin{pmatrix} [A^{-1}]^{-1} & 0 \\ 0 & 0 \end{pmatrix} = \begin{pmatrix} A_{12}A_{22}^{-1}A_{21} & A_{12} \\ A_{21} & A_{22} \end{pmatrix} \geq 0. \tag{6.10}$$

We end this section by a matrix inequality on principal submatrices. Since a principal submatrix of a positive definite matrix is also positive definite, we have, for any $n \times n$ matrices A, B, and C,

$$\begin{pmatrix} A & B \\ B^* & C \end{pmatrix} \geq 0 \Rightarrow \begin{pmatrix} [A] & [B] \\ [B^*] & [C] \end{pmatrix} \geq 0.$$

Using the partitioned matrix in (6.9), we obtain the following result.

Theorem 6.14 *Let A be an $n \times n$ positive definite matrix. Then for any $n \times n$ matrix B, with $[\,\cdot\,]$ standing for a principal submatrix,*

$$[B^*][A]^{-1}[B] \leq [B^*A^{-1}B].$$

Problems

1. Show that $\begin{pmatrix} A & A \\ A & X \end{pmatrix} \geq 0$ for any $X \geq A \geq 0$.

2. Show that if $A \geq B \geq 0$, then $\begin{pmatrix} A & B \\ B & A \end{pmatrix} \geq 0$.

3. Show that X must be the zero matrix if $\begin{pmatrix} I+X & I \\ I & I-X \end{pmatrix} \geq 0$.

4. Show that $\begin{pmatrix} I & X \\ X^* & X^*X \end{pmatrix} \geq 0$ for any matrix X.

5. Show that if $A \geq 0$ and $B \geq 0$ are of the same size, then

$$\begin{pmatrix} A & A^{\frac{1}{2}}B^{\frac{1}{2}} \\ B^{\frac{1}{2}}A^{\frac{1}{2}} & B \end{pmatrix} \geq 0.$$

    In particular, for any positive semidefinite matrix A

$$\begin{pmatrix} A & A^{\frac{1}{2}} \\ A^{\frac{1}{2}} & I \end{pmatrix} \geq 0.$$

6. Let $A > 0$ and $B > 0$ be of the same size. Show that

$$\begin{pmatrix} A+B & A \\ A & A+X \end{pmatrix} \geq 0 \;\Leftrightarrow\; X \geq -(A^{-1}+B^{-1})^{-1}.$$

7. Refer to Theorem 6.10. Show the reversal Fischer inequality

$$\det \widetilde{A_{11}} \det \widetilde{A_{22}} \leq \det A.$$

8. Show the Hadamard determinantal inequality by Theorem 6.3(4).

9. Let A be an $m \times n$ complex matrix and B be an $n \times n$ positive definite matrix. If $A^*BA = 0$, show that $A = 0$. Does the assertion hold if B is a nonzero positive definite or general nonsingular matrix?

10. When does equality in (6.8) occur?

11. Show that a square complex matrix A is unitary if and only if each row (column) vector of A has length 1 and $|\det A| = 1$.

12. Let A be $n \times n$ and B be $n \times m$. If A is nonsingular, verify that

$$\begin{pmatrix} A & B \\ B^* & B^*A^{-1}B \end{pmatrix} = \begin{pmatrix} I & 0 \\ 0 & B^* \end{pmatrix} \begin{pmatrix} A & I \\ I & A^{-1} \end{pmatrix} \begin{pmatrix} I & 0 \\ 0 & B \end{pmatrix}.$$

13. Show that the following matrices are positive semidefinite:

    (a) $\begin{pmatrix} \sigma_{\max}(A)I_n & A^* \\ A & \sigma_{\max}(A)I_m \end{pmatrix}$ for any $m \times n$ matrix A;

    (b) $\begin{pmatrix} A & I \\ I & A^{-1} \end{pmatrix}$ for any positive definite matrix A;

    (c) $\begin{pmatrix} A^*A & A^*B \\ B^*A & B^*B \end{pmatrix}$ for any A and B of the same size;

    (d) $\begin{pmatrix} I + A^*A & A^* + B^* \\ A + B & I + BB^* \end{pmatrix}$ for any A and B of the same size;

    (e) $\begin{pmatrix} \lambda A & A \\ A & \frac{1}{\lambda}A \end{pmatrix}$ for any $\lambda \in (0, +\infty)$ and $A \geq 0$.

14. Let $A > 0$ and $B > 0$ be of the same size. Show that, with $[X]$ standing for the corresponding principal submatrices of X,

    (a) $[(A+B)^{-1}] \leq [A^{-1}] + [B^{-1}]$,
    (b) $[A+B]^{-1} \leq [A]^{-1} + [B]^{-1}$,
    (c) $[A+B]^{-1} \leq [(A+B)^{-1}]$,
    (d) $[A]^{-1} + [B]^{-1} \leq [A^{-1}] + [B^{-1}]$.

15. Show that for any square complex matrix A, with $[X]$ standing for the corresponding principal submatrices of X,

$$[A^*A] \geq [A^*][A].$$

16. Let A and B be square complex matrices of the same size. With $[X]$ standing for the corresponding principal submatrices of X, show that

$$[AB][B^*A^*] \leq [ABB^*A^*]$$

and

$$[A][B][B^*][A^*] \leq [A][BB^*][A^*].$$

Show by example that the inequalities below do not hold in general:
$$[A][B][B^*][A^*] \leq [ABB^*A^*];$$
$$[AB][B^*A^*] \leq [A][BB^*][A^*];$$
$$[A][BB^*][A^*] \leq [ABB^*A^*].$$
Conclude that the following inequality does not hold in general:
$$[B^*][A][B] \leq [B^*AB], \quad \text{where } A \geq 0.$$

17. Let A be a positive semidefinite matrix. Show by the given A that
$$[A^4] \geq [A]^4$$
is not true in general, where $[\,\cdot\,]$ stands for a principal submatrix and
$$A = \begin{pmatrix} 1 & 0 & 1 \\ 0 & 0 & 1 \\ 1 & 1 & 1 \end{pmatrix}.$$
What is wrong with the proof, using $[A^2] \geq [A]^2$ in Theorem 6.12,
$$[A^4] = [(A^2)^2] \geq [A^2]^2 \geq ([A]^2)^2 = [A]^4?$$

18. Let $A > 0$ and $B > 0$ be of the same size. Show that
$$A - X^*B^{-1}X > 0 \quad \Leftrightarrow \quad B - XA^{-1}X^* > 0.$$

19. Let $A > 0$ and $B \geq 0$ be of the same size. Show that
$$\operatorname{tr}(A^{-1}B) \geq \frac{\operatorname{tr} B}{\lambda_{\max}(A)} \geq \frac{\operatorname{tr} B}{\operatorname{tr} A}.$$

20. Let $A \in \mathbb{M}_n$, $C \in \mathbb{M}_m$, and B be $n \times m$. Show that
$$\begin{pmatrix} A & B \\ B^* & C \end{pmatrix} \geq 0 \quad \Rightarrow \quad \operatorname{tr}(B^*B) \leq \operatorname{tr} A \operatorname{tr} C.$$
Does
$$\det(B^*B) \leq \det A \det C?$$

21. Let A, B, and C be n-square matrices. If $AB = BA$, show that
$$\begin{pmatrix} A & B \\ B^* & C \end{pmatrix} \geq 0 \quad \Rightarrow \quad A^{\frac{1}{2}}CA^{\frac{1}{2}} \geq B^*B.$$

22. Let A, B, and C be square matrices of the same size. Show that
$$\begin{pmatrix} A & B \\ B^* & C \end{pmatrix} \geq 0 \quad \Rightarrow \quad A + B + B^* + C \geq 0.$$

23. Let $A \in \mathbb{M}_n$ be a positive definite matrix. Show that for any $B \in \mathbb{M}_n$,
$$A + B + B^* + B^* A^{-1} B \geq 0.$$

In particular,
$$I + B + B^* + B^* B \geq 0.$$

24. Find a positive semidefinite matrix A partitioned as
$$A = \begin{pmatrix} A_{11} & A_{12} \\ A_{21} & A_{22} \end{pmatrix}$$
such that
$$\det A = \det A_{11} \det A_{22} \quad \text{but} \quad A_{12} = A_{21}^* \neq 0.$$

25. Show that for any complex matrices A and B of the same size,
$$\begin{pmatrix} \det(A^*A) & \det(A^*B) \\ \det(B^*A) & \det(B^*B) \end{pmatrix} \geq 0, \quad \begin{pmatrix} \operatorname{tr}(A^*A) & \operatorname{tr}(A^*B) \\ \operatorname{tr}(B^*A) & \operatorname{tr}(B^*B) \end{pmatrix} \geq 0,$$
and
$$\begin{vmatrix} A^*A & A^*B \\ B^*A & B^*B \end{vmatrix} = 0.$$

26. For any vectors $x_1, x_2, \ldots, x_n$ in an inner product space, let
$$G(x_1, x_2, \ldots, x_n) = ((x_i, x_j)).$$
Show that
$$G(x_1, x_2, \ldots, x_n) \geq 0$$
and that
$$\det G(x_1, \ldots, x_n) \leq \prod_{i=1}^{n} (x_i, x_i).$$
Equality holds if and only if the vectors are orthogonal. Moreover,
$$\det G(x_1, \ldots, x_n) \leq \det G(x_1, \ldots, x_m) \det G(x_{m+1}, \ldots, x_n).$$

27. Let $A = B + Ci \geq 0$, where B and C are real matrices. Show that
   (a) B is positive semidefinite,
   (b) C is skew-symmetric,
   (c) $\det B \geq \det A$,
   (d) $\text{rank}(B) \geq \max\{\text{rank}(A), \text{rank}(C)\}$.

   When does equality in (c) occur?

28. Let A be an $n \times n$ positive semidefinite matrix partitioned as
$$A = \begin{pmatrix} A_{11} & A_{12} \\ A_{21} & A_{22} \end{pmatrix}, \quad \text{where } A_{11} \text{ and } A_{22} \text{ are square}.$$

   Show that, by writing $A = X^*X$, where $X = (S, T)$ for some S, T,
$$\mathcal{C}(A_{12}) \subseteq \mathcal{C}(A_{11}), \quad \mathcal{C}(A_{21}) \subseteq \mathcal{C}(A_{22}),$$
   and
$$\mathcal{R}(A_{12}) \subseteq \mathcal{R}(A_{22}), \quad \mathcal{R}(A_{21}) \subseteq \mathcal{R}(A_{11}).$$

   Further show that
$$\text{rank}(A_{11}, A_{12}) = \text{rank}(A_{11}), \quad \text{rank}(A_{21}, A_{22}) = \text{rank}(A_{22}).$$

   Derive that $A_{12} = A_{11}P$ and $A_{21} = QA_{11}$ for some P and Q. Thus
$$\max\{\text{rank}(A_{12}), \text{rank}(A_{21})\} \leq \min\{\text{rank}(A_{11}), \text{rank}(A_{22})\}.$$

29. Let $[X]$ stand for a principal submatrix of X. If $A \geq 0$, show that
$$\text{rank}[A^k] = \text{rank}[A]^k = \text{rank}[A], \quad k = 1, 2, \ldots.$$
   (Hint: $\text{rank}[AB] \leq \text{rank}[A]$ and $\text{rank}[A^2] = \text{rank}[A]$ for $A, B \geq 0$.)

6.4 Schur Complements and Determinantal Inequalities

Making use of Schur complements (or type III elementary operations for partitioned matrices) has appeared to be an important technique in many matrix problems and applications in statistics.

As defined in the previous section, the Schur complement of the nonsingular principal submatrix A_{11} in the partitioned matrix

$$A = \begin{pmatrix} A_{11} & A_{12} \\ A_{21} & A_{22} \end{pmatrix}$$

is

$$\widetilde{A_{11}} = A_{22} - A_{21} A_{11}^{-1} A_{12}.$$

Note that if A is positive semidefinite and if A_{11} is nonsingular, then

$$A_{22} \geq \widetilde{A_{11}} \geq 0.$$

Theorem 6.15 *Let $A > 0$ be partitioned as above. Then*

$$A^{-1} = \begin{pmatrix} \widetilde{A_{22}}^{-1} & X \\ Y & \widetilde{A_{11}}^{-1} \end{pmatrix}, \tag{6.11}$$

where

$$X = -A_{11}^{-1} A_{12} \widetilde{A_{11}}^{-1} = -\widetilde{A_{22}}^{-1} A_{12} A_{22}^{-1}$$

and

$$Y = -A_{22}^{-1} A_{21} \widetilde{A_{22}}^{-1} = -\widetilde{A_{11}}^{-1} A_{21} A_{11}^{-1}.$$

The proof of this theorem follows from Theorem 2.4 immediately.

The inverse form (6.11) of A in terms of Schur complements is very useful. We demonstrate an application to obtain some determinantal inequalities, and we end this section with the Hua inequality.

Theorem 6.16 *For any n-square complex matrices A and B,*

$$\det(I + AA^*) \det(I + B^*B) \geq |\det(A + B)|^2 + |\det(I - AB^*)|^2$$

with equality if and only if $n = 1$ or $A + B = 0$ or $AB^* = I$.

Sec. 6.4 Schur Complements and Determinantal Inequalities

The determinantal inequality will follow from the following key matrix identity, for which we will present two proofs:

$$I + AA^* = (A+B)(I+B^*B)^{-1}(A+B)^* \\ + (I - AB^*)(I + BB^*)^{-1}(I - AB^*)^*. \quad (6.12)$$

Note that the left-hand side of (6.12) is independent of B.

PROOF 1 FOR THE IDENTITY (6.12). Use Schur complements. Let

$$X = \begin{pmatrix} I + B^*B & B^* + A^* \\ A + B & I + AA^* \end{pmatrix}.$$

Then the Schur complement of $I + B^*B$ in X is

$$(I + AA^*) - (A+B)(I+B^*B)^{-1}(A+B)^*. \quad (6.13)$$

On the other hand, we write

$$X = \begin{pmatrix} I & B^* \\ A & I \end{pmatrix} \begin{pmatrix} I & A^* \\ B & I \end{pmatrix}.$$

Then by using (6.11), if $I - AB^*$ is invertible (then so is $I - B^*A$),

$$X^{-1} = \begin{pmatrix} I & A^* \\ B & I \end{pmatrix}^{-1} \begin{pmatrix} I & B^* \\ A & I \end{pmatrix}^{-1}$$
$$= \begin{pmatrix} (I - A^*B)^{-1} & -(I - A^*B)^{-1}A^* \\ -(I - BA^*)^{-1}B & (I - BA^*)^{-1} \end{pmatrix}$$
$$\times \begin{pmatrix} (I - B^*A)^{-1} & -B^*(I - AB^*)^{-1} \\ -A(I - B^*A)^{-1} & (I - AB^*)^{-1} \end{pmatrix}.$$

Thus, we have the lower-right corner of X^{-1}, after multiplying out the right-hand side and then taking inverses,

$$(I - AB^*)(I + BB^*)^{-1}(I - AB^*)^*. \quad (6.14)$$

Equating (6.13) and (6.14), by (6.11), results in (6.12). The singular case of $I - AB^*$ follows from a continuity argument by replacing A with ϵA such that $I - \epsilon AB^*$ is invertible and by letting $\epsilon \to 1$.

PROOF 2 FOR THE IDENTITY (6.12). A direct proof by showing that

$$(I + AA^*) - (I - AB^*)(I + BB^*)^{-1}(I - AB^*)^*$$

equals

$$(A + B)(I + B^*B)^{-1}(A + B)^*.$$

Noticing that

$$B(I + B^*B) = (I + BB^*)B,$$

we have, by multiplying the inverses,

$$(I + BB^*)^{-1}B = B(I + B^*B)^{-1} \qquad (6.15)$$

and, by taking the conjugate transpose,

$$B^*(I + BB^*)^{-1} = (I + B^*B)^{-1}B^*. \qquad (6.16)$$

Furthermore, the identity

$$I = (I + B^*B)(I + B^*B)^{-1}$$

yields

$$I - B^*B(I + B^*B)^{-1} = (I + B^*B)^{-1} \qquad (6.17)$$

and, by switching B and B^*,

$$I - (I + BB^*)^{-1} = BB^*(I + BB^*)^{-1}. \qquad (6.18)$$

Upon computation, we have

$$\begin{aligned}
&(I + AA^*) - (I - AB^*)(I + BB^*)^{-1}(I - AB^*)^* \\
&= AA^* - AB^*(I + BB^*)^{-1}BA^* + AB^*(I + BB^*)^{-1} \\
&\quad + (I + BB^*)^{-1}BA^* + I - (I + BB^*)^{-1} \quad \text{(by expansion)} \\
&= AA^* - AB^*B(I + B^*B)^{-1}A^* + A(I + B^*B)^{-1}B^* \\
&\quad + B(I + B^*B)^{-1}A^* + BB^*(I + BB^*)^{-1} \quad \text{(by (6.15, 16, 18)} \\
&= A(I + B^*B)^{-1}A^* + A(I + B^*B)^{-1}B^* \\
&\quad + B(I + B^*B)^{-1}B^* + B(I + B^*B)^{-1}B^* \quad \text{(by (6.16, 17)} \\
&= (A + B)(I + B^*B)^{-1}(A + B)^* \quad \text{(by factoring)}.
\end{aligned}$$

SEC. 6.4 SCHUR COMPLEMENTS AND DETERMINANTAL INEQUALITIES

The identity (6.12) thus follows.

Now we are ready to prove the determinantal inequality. Recall that (Theorem 6.7 and Problem 28 of Section 6.2) for any positive semidefinite matrices X and Y of the same size (> 1),

$$\det(X+Y) \geq \det X + \det Y$$

with equality if and only if $X+Y$ is singular or $X = 0$ or $Y = 0$.

Applying this to (6.12) and noticing that $I + AA^*$ is never singular, we have, when A and B are square matrices of the same size,

$$|\det(I - AB^*)|^2 + |\det(A+B)|^2 \leq \det(I+AA^*)\det(I+B^*B);$$

equality holds if and only if $n = 1$ or $A + B = 0$ or $AB^* = I$. ∎

As consequences of (6.12), we have the Löwner partial orderings

$$I + AA^* \geq (A+B)(I+B^*B)^{-1}(A+B)^* \geq 0$$

and

$$I + AA^* \geq (I - AB^*)(I + BB^*)^{-1}(I - AB^*)^* \geq 0,$$

and thus the determinantal inequalities

$$|\det(A+B)|^2 \leq \det(I+AA^*)\det(I+B^*B)$$

and

$$|\det(I - AB^*)|^2 \leq \det(I+AA^*)\det(I+B^*B).$$

Using similar ideas, one proves the *Hua inequality* (Problem 14):

If A and B are $m \times n$ matrices such that $I - A^*A > 0$ and $I - B^*B > 0$, that is, A and B are strictly contractive matrices, then

$$(I - B^*A)(I - A^*A)^{-1}(I - A^*B) \geq I - B^*B,$$

or equivalently

$$\begin{pmatrix} (I-A^*A)^{-1} & (I-B^*A)^{-1} \\ (I-A^*B)^{-1} & (I-B^*B)^{-1} \end{pmatrix} \geq 0.$$

Consequently,

$$|\det(I - A^*B)|^2 \geq \det(I - A^*A)\det(I - B^*B).$$

Equality holds if and only if $A = B$.

Problems

1. Let A_{11} be a principal submatrix of a square matrix A. Show that
$$A > 0 \quad \Leftrightarrow \quad A_{11} > 0 \text{ and } \widetilde{A_{11}} > 0.$$

2. Show by writing $A = (A^{-1})^{-1}$ that for any principal submatrix A_{11},
$$\widetilde{\widetilde{A_{11}}^{-1}}^{-1} = A_{11} \quad \text{or} \quad \widetilde{\widetilde{A_{11}}^{-1}}^{-1} = A_{11}^{-1}.$$

3. Let A_{11} and B_{11} be the corresponding principal submatrices of the $n \times n$ positive semidefinite matrices A and B, respectively. Show that
$$4(A_{11} + B_{11})^{-1} \leq \widetilde{A_{22}}^{-1} + \widetilde{B_{22}}^{-1}.$$

4. Let A and B be positive definite. Show that $A \leq B \Rightarrow \widetilde{A_{11}} \leq \widetilde{B_{11}}$.

5. Let $A \geq 0$. Show that for any matrices X and Y of appropriate sizes,
$$\begin{pmatrix} X^*AX & X^*AY \\ Y^*AX & Y^*AY \end{pmatrix} \geq 0.$$

6. Use the block matrix $\begin{pmatrix} I & A \\ A^* & I \end{pmatrix}$ to show the matrix identities
$$(I - A^*A)^{-1} = I + A^*(I - AA^*)^{-1}A$$
and
$$I + AA^*(I - AA^*)^{-1} = (I - AA^*)^{-1}.$$

7. Show that if P is the elementary matrix of a type III operation on
$$A = \begin{pmatrix} A_{11} & A_{12} \\ A_{21} & A_{22} \end{pmatrix}, \quad \text{that is,} \quad P = \begin{pmatrix} I & X \\ 0 & I \end{pmatrix}.$$
then the Schur complements of A_{11} in A and in $P^T A P$ are the same.

8. Let $A, B, C,$ and D be $n \times n$ nonsingular complex matrices. Show that
$$\begin{vmatrix} A^{-1} & B^{-1} \\ C^{-1} & D^{-1} \end{vmatrix} = \frac{(-1)^n}{\det(ACBD)} \begin{vmatrix} A & C \\ B & D \end{vmatrix}.$$

9. Let A, C be $m \times n$ matrices, and B, D be $m \times p$ matrices. Show that
$$\begin{pmatrix} AA^* + BB^* & AC^* + BD^* \\ CA^* + DB^* & CC^* + DD^* \end{pmatrix} \geq 0.$$

10. Show that for matrices A, B, C, and D of appropriate sizes,
$$|\det(AC + BD)|^2 \le \det(AA^* + BB^*)\det(C^*C + D^*D).$$
In particular, for any two square matrices X and Y of the same size,
$$|\det(X + Y)|^2 \le \det(I + XX^*)\det(I + Y^*Y)$$
and
$$|\det(I + XY)|^2 \le \det(I + XX^*)\det(I + Y^*Y).$$

11. Prove or disprove that for any n-square complex matrices A and B,
$$\det(A^*A + B^*B) = \det(A^*A + BB^*)$$
or
$$\det(A^*A + B^*B) = \det(AA^* + BB^*).$$

12. Let A and B be $m \times n$ matrices. Show that for any $n \times m$ matrix X,
$$\begin{aligned}AA^* + BB^* &= (B + AX)(I + X^*X)^{-1}(B + AX)^* \\ &\quad + (A - BX^*)(I + XX^*)^{-1}(A - BX^*)^*.\end{aligned}$$

13. Let A and B be square contractive matrices of the same size. Derive
$$\det(I - A^*A)\det(I - B^*B) + |\det(A^* - B^*)|^2 \le |\det(I - A^*B)|^2$$
by applying Theorem 6.15 to the block matrix
$$\begin{pmatrix} I - A^*A & I - A^*B \\ I - B^*A & I - B^*B \end{pmatrix} = \begin{pmatrix} I & A^* \\ I & B^* \end{pmatrix} \begin{pmatrix} I & I \\ -A & -B \end{pmatrix}.$$
Show that the determinant of the block matrix on the left-hand side is $(-1)^n|\det(A - B)|^2$. As a consequence of the inequality,
$$|\det(I - A^*B)|^2 \ge \det(I - A^*A)\det(I - B^*B)$$
with equality if and only if $A = B$.

14. Show the Hua inequality by the method in the second proof of (6.12).

15. Let A, B, C, and D be square matrices of the same size. Show that
$$I + D^*C - (I + D^*B)(I + A^*B)^{-1}(I + A^*C)$$
$$= (D - A)^*(I + BA^*)^{-1}(C - B)$$
if the inverses involved exist, by considering the block matrix
$$\begin{pmatrix} I + A^*B & I + A^*C \\ I + D^*B & I + D^*C \end{pmatrix}.$$

6.5 The Kronecker Product and Hadamard Product

Matrices can be multiplied in different ways. The Kronecker product and Hadamard product, defined below, used in many fields, are almost as important as the ordinary product.

The *Kronecker product*, also known as *tensor product* or *direct product*, of two matrices A and B of sizes $m \times n$ and $s \times t$, respectively, is defined to be the $(ms) \times (nt)$ matrix

$$A \otimes B = \begin{pmatrix} a_{11}B & a_{12}B & \cdots & a_{1n}B \\ a_{21}B & a_{22}B & \cdots & a_{2n}B \\ \vdots & \vdots & \vdots & \vdots \\ a_{m1}B & a_{m2}B & \cdots & a_{mn}B \end{pmatrix}.$$

The *Hadamard product*, or the *Schur product*, of two matrices A and B of the same size is defined to be the entrywise product

$$A \circ B = (a_{ij}b_{ij}).$$

In particular, for $u = (u_1, u_2, \ldots, u_n)^T$, $v = (v_1, v_2, \ldots, v_n)^T \in \mathbb{C}^n$,

$$u \otimes v = (u_1 v_1, \ldots, u_1 v_n, \ldots, u_n v_1, \ldots, u_n v_n)^T$$

and

$$u \circ v = (u_1 v_1, u_2 v_2, \ldots, u_n v_n)^T.$$

The Kronecker product has the following basic properties, which are verified by definition and direct computations.

Theorem 6.17 *For matrices A, B, C, and D of appropriate sizes*

1. $(A \otimes B)(C \otimes D) = (AC) \otimes (BD)$,

2. $(A \otimes B)^* = A^* \otimes B^*$,

3. $(A \otimes B)^{-1} = A^{-1} \otimes B^{-1}$ *if A and B are invertible*,

4. $A \otimes B$ *is unitary if A and B are unitary.*

SEC. 6.5   THE KRONECKER PRODUCT AND HADAMARD PRODUCT   191

An interesting and important observation is that the Hadamard product $A \circ B$ is contained in the Kronecker product $A \otimes B$ as a principal submatrix if A and B are square matrices of the same size.

Theorem 6.18 *Let $A, B \in \mathbb{M}_n$. Then the Hadamard product $A \circ B$ is a principal submatrix of the Kronecker product $A \otimes B$ lying on the intersections of rows and columns $1, n+2, 2n+3, \ldots, n^2$.*

PROOF. Let e_i be, as usual, the column vector of n components with the ith position 1 and 0 elsewhere, $i = 1, 2, \ldots, n$, and let

$$E = (e_1 \otimes e_1, \ldots, e_n \otimes e_n).$$

Then for every pair of i and j, we have by computation

$$a_{ij} b_{ij} = (e_i^T A e_j) \otimes (e_i^T B e_j) = (e_i \otimes e_i)^T (A \otimes B)(e_j \otimes e_j),$$

which equals the (i, j)-entry of the matrix $E^T(A \otimes B)E$. Thus,

$$E^T(A \otimes B)E = A \circ B.$$

This says that $A \circ B$ is the principal submatrix of $A \otimes B$ lying on the intersections of rows and columns $1, n+2, 2n+3, \ldots, n^2$. ∎

The following theorem, relating the eigenvalues of the Kronecker product to those of individual matrices, presents in its proof a common method of decomposing a Kronecker product.

Theorem 6.19 *Let A and B be m-square and n-square complex matrices with eigenvalues λ_i and μ_j, $i = 1, \ldots, m$, $j = 1, \ldots, n$, respectively. Then the eigenvalues of $A \otimes B$ are*

$$\lambda_i \mu_j, \quad i = 1, \ldots, m, \ j = 1, \ldots, n,$$

and the eigenvalues of $A \otimes I_n + I_m \otimes B$ are

$$\lambda_i + \mu_j, \quad i = 1, \ldots, m, \ j = 1, \ldots, n.$$

PROOF. Let U and V be unitary matrices such that

$$U^* A U = T_1 \quad \text{and} \quad V^* B V = T_2,$$

where T_1 and T_2 are upper-triangular matrices with diagonal entries λ_i and μ_j, $i = 1, \ldots, m$, $j = 1, \ldots, n$, respectively. Then

$$T_1 \otimes T_2 = (U^*AU) \otimes (V^*BV) = (U^* \otimes V^*)(A \otimes B)(U \otimes V).$$

Note that $U \otimes V$ is unitary. Thus $A \otimes B$ is unitarily similar to $T_1 \otimes T_2$. The eigenvalues of the latter matrix are $\lambda_i \mu_j$.

For the second part, let $W = U \otimes V$. Then

$$W^*(A \otimes I_n)W = T_1 \otimes I_n = \begin{pmatrix} \lambda_1 I_n & & * \\ & \ddots & \\ 0 & & \lambda_m I_n \end{pmatrix}$$

and

$$W^*(I_m \otimes B)W = I_m \otimes T_2 = \begin{pmatrix} T_2 & & 0 \\ & \ddots & \\ 0 & & T_2 \end{pmatrix}.$$

Thus

$$W^*(A \otimes I_n + I_m \otimes B)W = T_1 \otimes I_n + I_m \otimes T_2$$

is an upper-triangular matrix with eigenvalues $\lambda_i + \mu_j$. ∎

It follows that $A > 0$, $B > 0$ $\Rightarrow$ $A \otimes B > 0$ and that the eigenvalues of $A \otimes I_m + I_m \otimes A$ are $\lambda_i + \lambda_j$, $i, j = 1, \ldots, m$.

The above theorem implies the following at once.

Theorem 6.20 *Let $A \geq 0$ and $B \geq 0$. Then $A \otimes B \geq 0$.*

Note that A and B in the theorem may have different sizes.

Our next celebrated theorem of Schur on Hadamard products will be used repeatedly in deriving matrix inequalities that involve Hadamard products of positive semidefinite matrices.

Theorem 6.21 (Schur) *Let A and B be n-square matrices. Then*

$$A \geq 0, \ B \geq 0 \quad \Rightarrow \quad A \circ B \geq 0$$

and

$$A > 0, \ B > 0 \quad \Rightarrow \quad A \circ B > 0.$$

PROOF 1. Since the Hadamard product $A \circ B$ is a principal submatrix of the Kronecker product $A \otimes B$, which is positive semidefinite by the preceding theorem, the positive semidefiniteness of $A \circ B$ follows.

For the positive definite case, it is sufficient to notice that a principal submatrix of a positive definite matrix is also positive definite.

PROOF 2. Write, by Theorem 6.3, $A = U^*U$ and $B = V^*V$, and let u_i and v_i be the ith columns of matrices U and V, respectively. Then

$$a_{ij} = u_i^* u_j = (u_j, u_i), \quad b_{ij} = v_i^* v_j = (v_j, v_i)$$

for each pair i and j, and thus (Problem 8)

$$A \circ B = (a_{ij} b_{ij}) = \Big((u_j, u_i)(v_j, v_i)\Big) = \Big((u_j \otimes v_j, u_i \otimes v_i)\Big) \geq 0.$$

PROOF 3. For vector $x \in \mathbb{C}^n$, denote by $\operatorname{diag} x$ the n-square diagonal matrix with the components of x on the diagonal. We have

$$\begin{aligned} x^*(A \circ B)x &= \operatorname{tr}(\operatorname{diag} x^* \; A \; \operatorname{diag} x \; B^T) \\ &= \operatorname{tr}\left((B^{\frac{1}{2}})^T \; \operatorname{diag} x^* \; A^{\frac{1}{2}} A^{\frac{1}{2}} \; \operatorname{diag} x \; (B^{\frac{1}{2}})^T\right) \\ &= \operatorname{tr}\left(A^{\frac{1}{2}} \; \operatorname{diag} x \; (B^{\frac{1}{2}})^T\right)^* \left(A^{\frac{1}{2}} \; \operatorname{diag} x \; (B^{\frac{1}{2}})^T\right) \geq 0. \; \blacksquare \end{aligned}$$

We are now interested in comparing the pairs involving squares and inverses such as $(A \circ B)^2$ and $A^2 \circ B^2$, $(A \circ B)^{-1}$ and $A^{-1} \circ B^{-1}$.

Theorem 6.22 *Let $A \geq 0$ and $B \geq 0$ be of the same size. Then*

$$A^2 \circ B^2 \geq (A \circ B)^2.$$

Moreover, if A and B are nonsingular, then

$$A^{-1} \circ B^{-1} \geq (A \circ B)^{-1}$$

and

$$A \circ A^{-1} \geq I.$$

PROOF. Let a_i and b_i be the ith columns of matrices A and B, respectively. It is easy to verify by a direct computation that

$$(AA^*) \circ (BB^*) = (A \circ B)(A^* \circ B^*) + \sum_{i \neq j}(a_i \circ b_j)(a_i^* \circ b_j^*).$$

It follows that for any matrices A and B of the same size

$$(AA^*) \circ (BB^*) \geq (A \circ B)(A^* \circ B^*).$$

In particular, if A and B are positive semidefinite, then

$$A^2 \circ B^2 \geq (A \circ B)^2.$$

If A and B are nonsingular matrices, then, by Theorem 6.17,

$$(A \otimes B)^{-1} = A^{-1} \otimes B^{-1}.$$

Noticing that $A \circ B$ and $A^{-1} \circ B^{-1}$ are principal submatrices of $A \otimes B$ and $A^{-1} \otimes B^{-1}$ in the same position, respectively, we have by Theorem 6.12, with $[X]$ representing a principal submatrix of X,

$$A^{-1} \circ B^{-1} = [(A \otimes B)^{-1}] \geq [A \otimes B]^{-1} = (A \circ B)^{-1}.$$

To show the last inequality, partition A and A^{-1} conformally as

$$A = \begin{pmatrix} a & \alpha \\ \alpha^* & A_1 \end{pmatrix} \quad \text{and} \quad A^{-1} = \begin{pmatrix} b & \beta \\ \beta^* & B_1 \end{pmatrix}.$$

By inequality (6.10) in Section 6.3,

$$A - \begin{pmatrix} \frac{1}{b} & 0 \\ 0 & 0 \end{pmatrix} \geq 0, \quad A^{-1} - \begin{pmatrix} 0 & 0 \\ 0 & A_1^{-1} \end{pmatrix} \geq 0,$$

and by Theorem 6.20,

$$\left(A - \begin{pmatrix} \frac{1}{b} & 0 \\ 0 & 0 \end{pmatrix} \right) \circ \left(A^{-1} - \begin{pmatrix} 0 & 0 \\ 0 & A_1^{-1} \end{pmatrix} \right) \geq 0,$$

which yields

$$A \circ A^{-1} \geq \begin{pmatrix} 1 & 0 \\ 0 & A_1 \circ A_1^{-1} \end{pmatrix}.$$

An induction hypothesis on $A_1 \circ A_1^{-1}$ gives $A \circ A^{-1} \geq I$. ∎

Problems

1. Let A, B, C, and D be complex matrices. Show that

    (a) $A \otimes (B \otimes C) = (A \otimes B) \otimes C$,
    (b) $(A \otimes B)^k = A^k \otimes B^k$,
    (c) $\text{tr}(A \otimes B) = \text{tr}\, A \, \text{tr}\, B$,
    (d) $\text{rank}\,(A \otimes B) = \text{rank}\,(A)\,\text{rank}\,(B)$,
    (e) $\det(A \otimes B) = (\det A)^n (\det B)^m$, if $A \in \mathbb{M}_m$ and $B \in \mathbb{M}_n$,
    (f) $A \otimes B = 0$ if and only if $A = 0$ or $B = 0$,
    (g) if $A \otimes B = C \otimes D \neq 0$, where A and C are of the same size, then $A = aC$ and $B = bD$ with $ab = 1$, and vice versa.

2. Let A and B be m- and n-square matrices, respectively. Show that

$$(A \otimes I_n)(I_m \otimes B) = A \otimes B = (I_m \otimes B)(A \otimes I_n).$$

3. Let $A \in \mathbb{M}_n$ have characteristic polynomial p. Show that

$$\det(A \otimes I + I \otimes A) = (-1)^n \det p(-A).$$

4. Show that there exists a permutation matrix P such that

$$P^{-1}(A \otimes B)P = B \otimes A.$$

5. Show that the Kronecker product of two Hadamard matrices is also a Hadamard matrix.

6. Let $A \geq 0$ and $B \geq 0$ be of the same size. Show that

    (a) $\text{tr}(AB) \leq \text{tr}(A \otimes B) \leq \frac{1}{4}(\text{tr}\, A + \text{tr}\, B)^2$,
    (b) $\text{tr}(A \circ B) \leq \frac{1}{2}\text{tr}(A \circ A + B \circ B)$,
    (c) $\text{tr}(A \otimes B) \leq \frac{1}{2}\text{tr}(A \otimes A + B \otimes B)$,
    (d) $\det(A \otimes B) \leq \frac{1}{2}(\det(A \otimes A) + \det(B \otimes B))$.

7. Show that $A \geq 0 \Leftrightarrow \text{tr}(A \circ B) \geq 0$ for all $B \geq 0$, where A, $B \in \mathbb{M}_n$.

8. Let x, y, u, $v \in \mathbb{C}^n$. Show that, with $(x, y) = y^*x$ as usual,

$$(x, y)(u, v) = (x \otimes u, y \otimes v).$$

9. If $A = (a_{ij}) \geq 0$, show that the matrix $(|a_{ij}|^2) \geq 0$.

10. Consider the vector space $\mathbb{M}_2$ over $\mathbb{C}$.

    (a) What is the dimension of $\mathbb{M}_2$?
    (b) Find a basis for $\mathbb{M}_2$.
    (c) For $A, B \in \mathbb{M}_2$, define
    $$\mathcal{L}(X) = AXB, \quad X \in \mathbb{M}_2.$$
    Show that $\mathcal{L}$ is a linear transformation on $\mathbb{M}_2$.
    (d) Show that if λ and μ are eigenvalues of A and B, respectively, then $\lambda\mu$ is an eigenvalue of $\mathcal{L}$.

11. Let $A, B \in \mathbb{M}_n$. Show that $A \circ I_n = \operatorname{diag}(a_{11}, \ldots, a_{nn})$ and that
    $$D_1(A \circ B)D_2 = (D_1AD_2) \circ B = A \circ (D_1BD_2)$$
    for any n-square diagonal matrices D_1 and D_2.

12. Let $A \geq 0$ and $B \geq 0$ be of the same size. Show that
    $$\operatorname{rank}(A \circ B) \leq \operatorname{rank}(A) \operatorname{rank}(B).$$
    Show further that if $A > 0$, then $\operatorname{rank}(A \circ B)$ is equal to the number of nonzero diagonal entries of the matrix B.

13. Let $A > 0$ and let λ be any eigenvalue of $A^{-1} \circ A$. Show that

    (a) $\lambda \geq 1$,
    (b) $\det(A^{-1} \circ A) \geq 1$,
    (c) $A^{-1} + A \geq 2I$,
    (d) $A^{-1} \circ A^{-1} \geq (A \circ A)^{-1}$.

14. Let A, B, C, and D be $n \times n$ positive semidefinite matrices. Show that
    $$A \geq B \;\Rightarrow\; A \circ C \geq B \circ C, \quad A \otimes C \geq B \otimes C$$
    and
    $$A \geq B, \; C \geq D \;\Rightarrow\; A \circ C \geq B \circ D, \quad A \otimes C \geq B \otimes D.$$

15. Let $A > 0$ and $B > 0$ be of the same size. Show that
    $$(A^{-1} \circ B^{-1})^{-1} \leq A \circ B \leq (A^2 \circ B^2)^{\frac{1}{2}}.$$

16. Prove or disprove that for $A \geq 0$ and $B \geq 0$ of the same size
$$A^3 \circ B^3 \geq (A \circ B)^3 \quad \text{or} \quad A^{\frac{1}{2}} \circ B^{\frac{1}{2}} \leq (A \circ B)^{\frac{1}{2}}.$$

17. Show that for any square matrices A and B of the same size
$$(A \circ B)(A^* \circ B^*) \leq (\sigma^2 I) \circ (AA^*),$$
where $\sigma = \sigma_{\max}(B)$ is the largest singular value of B.

18. Let A and B be complex matrices of the same size. Show that
$$\begin{pmatrix} (AA^*) \circ I & A \circ B \\ A^* \circ B^* & (B^*B) \circ I \end{pmatrix} \geq 0, \quad \begin{pmatrix} (AA^*) \circ (BB^*) & A \circ B \\ A^* \circ B^* & I \end{pmatrix} \geq 0.$$

19. Let $A \geq 0$ and $B \geq 0$ be of the same size. Let λ be the largest eigenvalue of A and μ be the largest diagonal entry of B. Show that
$$A \circ B \leq \lambda I \circ B \leq \lambda \mu I$$
and
$$\begin{pmatrix} \lambda \mu I & A \circ B \\ A \circ B & \lambda \mu I \end{pmatrix} \geq 0.$$

20. Let A, B, C, D, X, Y, U, and V be $n \times n$ complex matrices and let
$$M = \begin{pmatrix} A & B \\ C & D \end{pmatrix}, \quad N = \begin{pmatrix} X & Y \\ U & V \end{pmatrix}.$$
Define
$$M \odot N = \begin{pmatrix} A \otimes X & B \otimes Y \\ C \otimes U & D \otimes V \end{pmatrix}.$$
Show that $M \circ N$ is a principal submatrix of $M \odot N$ and that $M \odot N$ is a principal submatrix of $M \otimes N$. Further $M \odot N \geq 0$ if $M, N \geq 0$.

21. Let $\lambda_1, \lambda_2, \ldots, \lambda_n$ be positive numbers. Use Cauchy matrices to show that the following matrices are positive semidefinite:
$$\left(\frac{1}{\lambda_i + \lambda_j}\right), \quad \left(\frac{1}{\lambda_i \lambda_j}\right), \quad \left(\frac{\lambda_i \lambda_j}{\lambda_i + \lambda_j}\right),$$
$$\left(\frac{1}{\lambda_i^2 + \lambda_j^2}\right), \quad \left(\sqrt{\lambda_i \lambda_j}\right), \quad \left(\frac{2}{\lambda_i^{-1} + \lambda_j^{-1}}\right),$$
$$\left(\frac{1}{\lambda_i(\lambda_i + \lambda_j)\lambda_j}\right), \quad \left(\frac{\sqrt{\lambda_i \lambda_j}}{\lambda_i + \lambda_j}\right), \quad \left(\frac{\lambda_i \lambda_j}{\sqrt{\lambda_i} + \sqrt{\lambda_j}}\right).$$

6.6 Schur Complements and Hadamard Products

The goal of this section is to obtain some inequalities for matrix sums and Hadamard products using Schur complements.

As we saw earlier (Theorem 6.7 and Theorem 6.22), for any positive definite matrices A and B of the same size

$$(A+B)^{-1} \leq \frac{1}{4}(A^{-1}+B^{-1})$$

and

$$(A \circ B)^{-1} \leq A^{-1} \circ B^{-1}.$$

These are special cases of the next theorem whose proof uses the fact

$$\begin{pmatrix} A & B \\ B^* & B^*A^{-1}B \end{pmatrix} \geq 0 \quad \text{if } A > 0.$$

Theorem 6.23 *Let A and B be n-square positive definite matrices, and let C and D be any matrices of size $m \times n$. Then*

$$(C+D)(A+B)^{-1}(C+D)^* \leq CA^{-1}C^* + DB^{-1}D^*, \qquad (6.19)$$

$$(C \circ D)(A \circ B)^{-1}(C \circ D)^* \leq (CA^{-1}C^*) \circ (DB^{-1}D^*). \qquad (6.20)$$

PROOF. Note that $X \geq 0$, $Y \geq 0$, $X+Y \geq 0$, and $X \circ Y \geq 0$, where

$$X = \begin{pmatrix} A & C^* \\ C & CA^{-1}C^* \end{pmatrix}, \quad Y = \begin{pmatrix} B & D^* \\ D & DB^{-1}D^* \end{pmatrix}.$$

The inequalities are immediate by taking the Schur complement of the (1, 1)-block in $X+Y \geq 0$ and $X \circ Y \geq 0$, respectively. ∎

An alternative approach to prove (6.20) is to use Theorem 6.14 with the observation that $X \circ Y$ is a principal submatrix of $X \otimes Y$.

By taking $A = B = I_n$ in (6.19) and (6.20), we have

$$\frac{1}{2}(C+D)(C+D)^* \leq CC^* + DD^*$$

and

$$(C \circ D)(C \circ D)^* \leq (CC^*) \circ (DD^*).$$

Theorem 6.24 *Let A and B be positive definite matrices of the same size partitioned conformally as*

$$A = \begin{pmatrix} A_{11} & A_{12} \\ A_{21} & A_{22} \end{pmatrix}, \quad B = \begin{pmatrix} B_{11} & B_{12} \\ B_{21} & B_{22} \end{pmatrix}.$$

Then
$$\widetilde{A_{11} + B_{11}} \geq \widetilde{A_{11}} + \widetilde{B_{11}} \tag{6.21}$$

and
$$\widetilde{A_{11} \circ B_{11}} \geq \widetilde{A_{11}} \circ \widetilde{B_{11}}. \tag{6.22}$$

PROOF. Let

$$\hat{A} = \begin{pmatrix} A_{11} & A_{12} \\ A_{21} & A_{21}A_{11}^{-1}A_{12} \end{pmatrix}, \quad \hat{B} = \begin{pmatrix} B_{11} & B_{12} \\ B_{21} & B_{21}B_{11}^{-1}B_{12} \end{pmatrix}.$$

The inequality (6.21) is obtained by taking the Schur complement of $A_{11} + B_{11}$ in $\hat{A} + \hat{B}$ and using (6.19). For (6.22), notice that

$$A_{22} \geq A_{21}A_{11}^{-1}A_{12}, \quad B_{22} \geq B_{21}B_{11}^{-1}B_{12}.$$

Therefore,

$$A_{22} \circ (B_{21}B_{11}^{-1}B_{12}) + B_{22} \circ (A_{21}A_{11}^{-1}A_{12})$$
$$\geq 2\Big((A_{21}A_{11}^{-1}A_{12}) \circ (B_{21}B_{11}^{-1}B_{12})\Big).$$

It follows that

$$\begin{aligned}\widetilde{A_{11}} \circ \widetilde{B_{11}} &= (A_{22} - A_{21}A_{11}^{-1}A_{12}) \circ (B_{22} - B_{21}B_{11}^{-1}B_{12}) \\ &\leq A_{22} \circ B_{22} - (A_{21}A_{11}^{-1}A_{12}) \circ (B_{21}B_{11}^{-1}B_{12}).\end{aligned}$$

Applying (6.20) to the right-hand side of the above inequality yields

$$\begin{aligned} & A_{22} \circ B_{22} - (A_{21}A_{11}^{-1}A_{12}) \circ (B_{21}B_{11}^{-1}B_{12}) \\ & \leq A_{22} \circ B_{22} - (A_{21} \circ B_{21})(A_{11} \circ B_{11})^{-1}(A_{12} \circ B_{12}) \\ & = \widetilde{A_{11} \circ B_{11}}.\end{aligned}$$

Thus
$$\widetilde{A_{11}} \circ \widetilde{B_{11}} \leq \widetilde{A_{11} \circ B_{11}}. \quad \blacksquare$$

We end this section with a determinantal inequality of Oppenheim. The proof uses a Schur complement technique.

As we recall, for $A \geq 0$ and any principal submatrix A_{11} of A

$$\det A = \det A_{11} \det \widetilde{A_{11}}.$$

Theorem 6.25 (Oppenheim) *Let A and B be $n \times n$ positive semidefinite matrices with diagonal entries a_{ii} and b_{ii}, respectively. Then*

$$\prod_{i=1}^{n} a_{ii} b_{ii} \geq \det(A \circ B) \geq a_{11} \cdots a_{nn} \det B \geq \det A \det B.$$

PROOF. The first and last inequalities are immediate from the Hadamard determinantal inequality. We show the second inequality.

Let $\hat{B}$ be as in the proof of the preceding theorem. Consider $A \circ \hat{B}$ this time and use induction on n, the order of the matrices.

If $n = 2$, then it is obvious. Suppose $n > 2$. Notice that

$$B_{21} B_{11}^{-1} B_{12} = B_{22} - \widetilde{B_{11}}.$$

Take the Schur complement of $A_{11} \circ B_{11}$ in $A \circ \hat{B}$ to get

$$A_{22} \circ (B_{22} - \widetilde{B_{11}}) - (A_{21} \circ B_{21})(A_{11} \circ B_{11})^{-1}(A_{12} \circ B_{12}) \geq 0$$

or

$$A_{22} \circ B_{22} - (A_{21} \circ B_{21})(A_{11} \circ B_{11})^{-1}(A_{12} \circ B_{12}) \geq A_{22} \circ \widetilde{B_{11}}.$$

Observe that the left-hand side of the above inequality is the Schur complement of $A_{11} \circ B_{11}$ in $A \circ B$. By taking determinants, we have

$$\det(\widetilde{A_{11} \circ B_{11}}) \geq \det(A_{22} \circ \widetilde{B_{11}}).$$

Multiply both sides by $\det(A_{11} \circ B_{11})$ to obtain

$$\det(A \circ B) \geq \det(A_{11} \circ B_{11}) \det(A_{22} \circ \widetilde{B_{11}}).$$

The assertion then follows from the induction hypothesis on the two determinants on the right-hand side. ■

The inequalities in the theorem may be rewritten in terms of eigenvalues as

$$\prod_{i=1}^{n} a_{ii} b_{ii} \geq \prod_{i=1}^{n} \lambda_i(A \circ B) \geq \prod_{i=1}^{n} a_{ii} \lambda_i(B) \geq \prod_{i=1}^{n} \lambda_i(AB) = \prod_{i=1}^{n} \lambda_i(A) \lambda_i(B).$$

Problems

1. Let $A > 0$. Use the identity $\det \widetilde{A_{11}} = \frac{\det A}{\det A_{11}}$ to show that

$$\frac{\det(A+B)}{\det(A_{11}+B_{11})} \geq \frac{\det A}{\det A_{11}} + \frac{\det B}{\det B_{11}}.$$

2. Let A, B, and $A+B$ be invertible matrices. Find the inverse of

$$\begin{pmatrix} A & A \\ A & A+B \end{pmatrix}.$$

Use the Schur complement of $A+B$ to verify that

$$A - A(A+B)^{-1}A = (A^{-1} + B^{-1})^{-1}.$$

3. Let $A > 0$ and $B > 0$ be of the same size. Show that for all $x, y \in \mathbb{C}^n$

$$(x+y)^*(A+B)^{-1}(x+y) \leq x^* A^{-1} x + y^* B^{-1} y$$

and

$$(x \circ y)^*(A \circ B)^{-1}(x \circ y) \leq (x^* A^{-1} x) \circ (y^* B^{-1} y).$$

4. Show that for any $m \times n$ complex matrices A and B

$$\begin{pmatrix} A^*A & A^* \\ A & I_m \end{pmatrix} \geq 0, \quad \begin{pmatrix} B^*B & B^* \\ B & I_m \end{pmatrix} \geq 0.$$

Derive the following inequalities using the Schur complement:

$$I_m \geq A(A^*A)^{-1}A^*, \quad \text{if } \text{rank}(A) = n;$$

$$(A^*A) \circ (B^*B) \geq (A^* \circ B^*)(A \circ B);$$

$$A^*A + B^*B \geq \frac{(A^* + B^*)(A+B)}{2}.$$

5. Let $A \in \mathbb{M}_n$ be a positive definite matrix. Show that for any $B \in \mathbb{M}_n$

$$\begin{pmatrix} A \circ (B^*A^{-1}B) & B^* \circ B \\ B^* \circ B & A \circ (B^*A^{-1}B) \end{pmatrix} \geq 0.$$

In particular,

$$\begin{pmatrix} A \circ A^{-1} & I \\ I & A \circ A^{-1} \end{pmatrix} \geq 0.$$

Derive the following inequalities using the Schur complement:

$$(A \circ A^{-1})^{-1} \leq A \circ A^{-1};$$

$$\det(B^* \circ B) \leq \det\left(A \circ (B^*A^{-1}B)\right);$$

$$\left(\operatorname{tr}(B^* \circ B)^2\right)^{\frac{1}{2}} \leq \operatorname{tr}\left(A \circ (B^*A^{-1}B)\right);$$

$$I \circ B^*B \geq (B^* \circ B)(I \circ B^*B)^{-1}(B \circ B)$$

if B has no zero row or column. Discuss the analog for sum.

6. Let A, B, and C be n-square complex matrices such that

$$\begin{pmatrix} A & B \\ B^* & C \end{pmatrix} \geq 0.$$

Show that

$$\left(\operatorname{tr}(B^* \star B)^2\right)^{\frac{1}{2}} \leq \operatorname{tr}(A \star C)$$

and

$$\det(B^* \star B) \leq \det(A \star C),$$

where $\star$ is $+$ or $\circ$.

7. Let A be a positive definite matrix partitioned as

$$A = \begin{pmatrix} A_{11} & A_{12} \\ A_{21} & A_{22} \end{pmatrix},$$

where A_{11} and A_{22} are square. Show that

$$A \circ A^{-1} \geq \begin{pmatrix} A_{11} \circ A_{11}^{-1} & 0 \\ 0 & \widetilde{A_{11}} \circ \widetilde{A_{11}}^{-1} \end{pmatrix} \geq 0.$$

Conclude that $A \circ A^{-1} \geq I$. Similarly show that $A^T \circ A^{-1} \geq I$.

⊙

6.7 The Cauchy-Schwarz and Kantorovich Inequalities

The Cauchy-Schwarz inequality is one of the most useful and fundamental inequalities in mathematics. It states that for any vectors x and y in an inner product vector space

$$|(x, y)|^2 \leq (x, x)(y, y)$$

with equality if and only if x and y are linearly dependent. Thus

$$|y^*x|^2 \leq (x^*x)(y^*y), \quad x, y \in \mathbb{C}^n.$$

We first give a Cauchy-Schwarz inequality involving block positive definite matrices, we then present the Kantorovich inequality.

Theorem 6.26 *Let A, B, and C be n-square matrices such that*

$$\begin{pmatrix} A & B^* \\ B & C \end{pmatrix} \geq 0.$$

Then

$$|(Bx, y)|^2 \leq (Ax, x)(Cy, y), \quad x, y \in \mathbb{C}^n.$$

PROOF. It is sufficient to notice that

$$\begin{pmatrix} x^* & 0 \\ 0 & y^* \end{pmatrix} \begin{pmatrix} A & B^* \\ B & C \end{pmatrix} \begin{pmatrix} x & 0 \\ 0 & y \end{pmatrix} = \begin{pmatrix} x^*Ax & x^*B^*y \\ y^*Bx & y^*Cy \end{pmatrix} \geq 0. \blacksquare$$

An alternative proof is to write $B = C^{\frac{1}{2}} X A^{\frac{1}{2}}$ for some contraction X, by Theorem 5.8, and then use the Cauchy-Schwarz inequality.

If A is an n-square positive definite matrix, with $A^{\frac{1}{2}}x$ and $A^{-\frac{1}{2}}y$ substituting for x and y in the Cauchy-Schwarz inequality, one gets

$$|y^*x|^2 \leq (x^*Ax)(y^*A^{-1}y).$$

In particular,

$$1 \leq (x^*Ax)(x^*A^{-1}x), \quad x^*x = 1.$$

We now show the *Kantorovich inequality*, a reversal of this.

If $A \in \mathbb{M}_n$ has real eigenvalues λ_i, we shall arrange the eigenvalues in the decreasing order $\lambda_{\max} = \lambda_1 \geq \lambda_2 \geq \cdots \geq \lambda_n = \lambda_{\min}$.

Theorem 6.27 (Kantorovich) *If $A \in \mathbb{M}_n$ is positive definite, then*

$$(x^*Ax)(x^*A^{-1}x) \leq \frac{(\lambda_1 + \lambda_n)^2}{4\lambda_1\lambda_n}, \quad x^*x = 1. \tag{6.23}$$

PROOF. We may assume that $A = \mathrm{diag}(\lambda_1, \lambda_2, \ldots, \lambda_n)$, since the inequality holds when A is replaced by U^*AU for any $n \times n$ unitary matrix U. Thus, (6.23) reduces to

$$\left(\sum_{i=1}^n \lambda_i |x_i|^2\right)\left(\sum_{i=1}^n \frac{1}{\lambda_i}|x_i|^2\right) \leq \frac{(\lambda_1 + \lambda_n)^2}{4\lambda_1\lambda_n}.$$

Equivalently,

$$\left(\sum_{i=1}^n t_i \lambda_i\right)\left(\sum_{i=1}^n t_i \lambda_i^{-1}\right) \leq a^2 g^{-2},$$

where $a = \frac{1}{2}(\lambda_1 + \lambda_n)$, $g = (\lambda_1\lambda_n)^{\frac{1}{2}}$, and $t_i \geq 0$ with $\sum_{i=1}^n t_i = 1$.

We may assume that $g = 1$ by replacing each λ_i by a multiple $\alpha\lambda_i$. Thus, $\lambda_1 = \frac{1}{\lambda_n}$ and each λ between λ_n and λ_1 satisfies

$$\lambda + \frac{1}{\lambda} \leq \lambda_n + \frac{1}{\lambda_n}.$$

It follows that

$$\sum_{i=1}^n t_i\lambda_i + \sum_{i=1}^n t_i\lambda_i^{-1} \leq \lambda_n + \frac{1}{\lambda_n} = 2a.$$

An application of the arithmetic mean-geometric mean inequality gives the desired inequality. ∎

The Kantorovich inequality has made appearances in a variety of forms. A matrix version is as follows: Let $A \in \mathbb{M}_n$ be a positive definite matrix. Then for any $n \times m$ matrix X satisfying $X^*X = I_m$

$$(X^*AX)^{-1} \leq X^*A^{-1}X \leq \frac{(\lambda_1 + \lambda_n)^2}{4\lambda_1\lambda_n}(X^*AX)^{-1}.$$

The first inequality is proven by noting that $I - Y(Y^*Y)^{-1}Y^* \geq 0$ for any matrix Y with columns linearly independent, while the second one needs more work by using the inequality: Let $0 < a < b$. Then

$$\frac{1}{x} \leq \frac{a+b}{ab} - \frac{x}{ab}, \quad \text{for any } x \in [a, b].$$

Choosing appropriate X (see the proof of Theorem 6.18) and replacing A with $A \otimes B$, one gets the inequalities on the Hadamard product of positive definite matrices

$$(A \circ B)^{-1} \leq A^{-1} \circ B^{-1} \leq \frac{(\lambda + \mu)^2}{4\lambda\mu}(A \circ B)^{-1},$$

where λ is the largest and μ is the smallest eigenvalue of $A \otimes B$. We leave the details to the reader (Problems 10 and 11).

Problems

1. Let A be an $n \times n$ positive definite matrix. Show that

$$|y^*x|^2 = x^*Ax\, y^*A^{-1}y$$

if and only if $y = 0$ or $Ax = cy$ for some constant c.

2. Let $\lambda_1, \ldots, \lambda_n$ be positive numbers. Show that for any $x, y \in \mathbb{C}^n$

$$\left|\sum_{i=1}^{n} x_i \overline{y_i}\right|^2 \leq \left(\sum_{i=1}^{n} \lambda_i |x_i|^2\right)\left(\sum_{i=1}^{n} \lambda_i^{-1} |y_i|^2\right).$$

3. Show that for positive numbers $a_1, a_2, \ldots, a_n$ and any $t \in [0, 1]$

$$\left(\sum_{i=1}^{n} a_i^{\frac{1}{2}}\right)^2 \leq \left(\sum_{i=1}^{n} a_i^t\right)\left(\sum_{i=1}^{n} a_i^{1-t}\right).$$

Equality occurs if and only if $t = \frac{1}{2}$ or all the a_i are equal.

4. Let $A \in \mathbb{M}_n$ be positive semidefinite. Show that for any unit $x \in \mathbb{C}^n$

$$(Ax, x)^2 \leq (A^2 x, x).$$

5. Let A, B, and C be $n \times n$ matrices. Assume that A and C are positive definite. Show that the following statements are equivalent:

(a) $\begin{pmatrix} A & B^* \\ B & C \end{pmatrix} \geq 0$;

(b) $\lambda_{\max}(BA^{-1}B^*C^{-1}) \leq 1$;

(c) $|(Bx, y)| \leq \frac{1}{2}((Ax, x) + (Cy, y))$ for all $x, y \in \mathbb{C}^n$.

6. Show that for any $n \times n$ matrix $A \geq 0$, $m \times n$ matrix B, $x, y \in \mathbb{C}^n$
$$|(Bx, y)|^2 \leq (Ax, x)(BA^{-1}B^*y, y)$$
and that for any $m \times n$ matrices A, B, and $x \in \mathbb{C}^n$, $y \in \mathbb{C}^m$
$$|((A + B)x, y)|^2 \leq ((I + A^*A)x, x)((I + BB^*)y, y).$$

7. Let A be an $n \times n$ positive definite matrix. Show that for all $x \neq 0$
$$1 \leq \frac{(x^*Ax)(x^*A^{-1}x)}{x^*x} \leq \frac{\lambda_1}{\lambda_n}.$$

8. Let A and B be $n \times n$ Hermitian. If $A \leq B$ or $B \leq A$, show that
$$A \circ B \leq \frac{1}{2}(A \circ A + B \circ B).$$

9. Let A and B be $n \times n$ positive definite matrices with eigenvalues contained in $[m, M]$, where $0 < m < M$. Show that for any $t \in [0, 1]$
$$tA^2 + (1-t)B^2 - (tA + (1-t)B)^2 \leq \frac{1}{4}(M - m)^2 I.$$

10. Show the Kantorovich inequality following the line:
    (a) if $0 < m \leq t \leq M$, then $0 \leq (m + M - t)t - mM$;
    (b) if $0 \leq m \leq \lambda_i \leq M$, $i = 1, \ldots, n$, and $\sum_{i=1}^{n} |x_i|^2 = 1$, then
    $$S\left(\frac{1}{\lambda}\right) \leq \frac{m + M - S(\lambda)}{mM},$$
    where $S(\frac{1}{\lambda}) = \sum_{i=1}^{n} \frac{1}{\lambda_i}|x_i|^2$ and $S(\lambda) = \sum_{i=1}^{n} \lambda_i |x_i|^2$;
    (c) the Kantorovich inequality follows from the inequality
    $$S(\lambda)S\left(\frac{1}{\lambda}\right) \leq \frac{(m + M)^2}{4mM}.$$

11. Let A be an $n \times n$ positive definite matrix. Show that for all $n \times m$ matrices X satisfying $X^*X = I_m$,
$$(X^*AX)^{-1} \leq X^*A^{-1}X \leq \frac{(\lambda_1 + \lambda_n)^2}{4\lambda_1\lambda_n}(X^*AX)^{-1}.$$
Use the first one to derive the inequality for principal submatrices:
$$[A]^{-1} \leq [A^{-1}]$$
and then show the following matrix inequalities:

(a) $X^*AX - (X^*A^{-1}X)^{-1} \le (\sqrt{\lambda_1} - \sqrt{\lambda_n})^2 I$,

(b) $(X^*AX)^2 \le X^*A^2X \le \frac{(\lambda_1+\lambda_n)^2}{4\lambda_1\lambda_n}(X^*AX)^2$,

(c) $X^*A^2X - (X^*AX)^2 \le \frac{(\lambda_1+\lambda_n)^2}{4} I$,

(d) $X^*AX \le (X^*A^2X)^{\frac{1}{2}} \le \frac{\lambda_1+\lambda_n}{2\sqrt{\lambda_1\lambda_n}}(X^*AX)$,

(e) $(X^*A^2X)^{\frac{1}{2}} - X^*AX \le \frac{(\lambda_1-\lambda_n)^2}{4(\lambda_1+\lambda_n)} I$.

12. Let A and B be $n \times n$ positive definite matrices. Show that
$$\lambda_{\max}(A \otimes B) = \lambda_{\max}(A)\lambda_{\max}(B) \quad \text{(denoted by } \lambda\text{)}$$
and
$$\lambda_{\min}(A \otimes B) = \lambda_{\min}(A)\lambda_{\min}(B) \quad \text{(denoted by } \mu\text{)}.$$
Derive the following inequalities from the previous problem:

(a) $(A \circ B)^{-1} \le A^{-1} \circ B^{-1} \le \frac{(\lambda+\mu)^2}{4\lambda\mu}(A \circ B)^{-1}$,

(b) $A \circ B - (A^{-1} \circ B^{-1})^{-1} \le (\sqrt{\lambda} - \sqrt{\mu})^2 I$,

(c) $(A \circ B)^2 \le A^2 \circ B^2 \le \frac{(\lambda+\mu)^2}{4\lambda\mu}(A \circ B)^2$,

(d) $(A \circ B)^2 - A^2 \circ B^2 \le \frac{(\lambda-\mu)^2}{4} I$,

(e) $A \circ B \le (A^2 \circ B^2)^{\frac{1}{2}} \le \frac{\lambda+\mu}{2\sqrt{\lambda\mu}} A \circ B$,

(f) $(A^2 \circ B^2)^{\frac{1}{2}} - A \circ B \le \frac{(\lambda-\mu)^2}{4(\lambda+\mu)} I$.

13. Let A be an $n \times n$ positive definite matrix. Show that
$$I \le A \circ A^{-1} \le \frac{\lambda_1^2 + \lambda_n^2}{4\lambda_1\lambda_n} I.$$

14. Let $\lambda_1 \ge \lambda_2 \ge \cdots \ge \lambda_n > 0$. Show that
$$\max_{i,j} \frac{(\lambda_i + \lambda_j)^2}{4\lambda_i\lambda_j} = \frac{(\lambda_1 + \lambda_n)^2}{4\lambda_1\lambda_n}.$$

15. (**Wielandt**) If A is an $n \times n$ positive semidefinite matrix, show that
$$|y^*Ax|^2 \le \left(\frac{\lambda_1 - \lambda_n}{\lambda_1 + \lambda_n}\right)^2 (x^*Ax)(y^*Ay)$$
for all orthogonal x and y in $\mathbb{C}^n$. (Hint: Consider $(A(x+ty),(x+ty))$.)

CHAPTER 7

Hermitian Matrices

INTRODUCTION This chapter contains fundamental results of Hermitian matrices and demonstrates the basic techniques used to derive the results. Section 7.1 presents equivalent conditions to matrix Hermitity, Section 7.2 gives some trace inequalities and discusses a necessary and sufficient condition for a square matrix to be a product of two Hermitian matrices, and Section 7.3 develops the min-max theorem and the interlacing theorem for eigenvalues. Section 7.4 deals with the eigenvalue and singular value inequalities for the sum of Hermitian matrices, and Section 7.5 shows a matrix triangle inequality.

7.1 Hermitian Matrices

A square complex matrix A is said to be *Hermitian* if A is equal to its transpose conjugate, symbolically, $A^* = A$.

Theorem 7.1 *An n-square complex matrix A is Hermitian if and only if there exists a unitary matrix U such that*

$$A = U^* \operatorname{diag}(\lambda_1, \ldots, \lambda_n)U, \qquad (7.1)$$

where the λ_i are real numbers.

SEC. 7.1　　　　　　　HERMITIAN MATRICES　　　　　　　209

In other words, A is Hermitian if and only if A is unitarily similar to a real diagonal matrix. This is the Hermitian case of the spectral decomposition theorem (Theorem 3.4). The decomposition (7.1) is often used when a trace or norm inequality is under investigation.

Theorem 7.2 *The following statements for $A \in \mathbb{M}_n$ are equivalent:*

1. *A is Hermitian;*

2. *$x^*Ax \in \mathbb{R}$ for all $x \in \mathbb{C}^n$;*

3. *$A^2 = A^*A$;*

4. *$\operatorname{tr} A^2 = \operatorname{tr}(A^*A)$.*

We show that (1)⇔(2) and (1)⇔(3). (1)⇔(4) is similar. It is not difficult to see that (1) and (2) are equivalent, since a complex number a is real if and only if $a^* = a$ and (Problem 14)

$$A^* = A \quad \Leftrightarrow \quad x^*(A^* - A)x = 0 \text{ for all } x \in \mathbb{C}^n.$$

We present four different proofs for (3)⇒(1), each of which shows a common technique of linear algebra and matrix theory. The first proof gives (4)⇒(1) immediately. Other implications are obvious.

PROOF 1. Use Schur decomposition. Write $A = U^*TU$, where U is unitary and T is upper-triangular with the eigenvalues $\lambda_1, \ldots, \lambda_n$ of A on the diagonal. Then $A^2 = A^*A$ implies $T^2 = T^*T$.

By comparison of the diagonal entries of the matrices on both sides of $T^2 = T^*T$, we have for each $j = 1, \ldots, n$,

$$\lambda_j^2 = |\lambda_j|^2 + \sum_{i<j} |t_{ij}|^2.$$

It follows that each λ_j is real and that $t_{ij} = 0$ whenever $i < j$. Therefore, T is real diagonal, and thus

$$A^* = (U^*TU)^* = U^*T^*U = U^*TU = A.$$

The trace identity $\operatorname{tr} T^2 = \operatorname{tr}(T^*T)$ yields (4)⇒(1) in the same way.

PROOF 2. Use the fact that $\operatorname{tr}(XX^*) = 0 \Leftrightarrow X = 0$. We show that $\operatorname{tr}(A - A^*)(A - A^*)^* = 0$ to conclude that $A - A^* = 0$ or $A = A^*$.

Upon computation we have

$$(A - A^*)(A - A^*)^* = AA^* - A^2 + A^*A - (A^*)^2,$$

which gives, by using $\operatorname{tr}(AA^*) = \operatorname{tr}(A^*A)$ and $A^*A = A^2 = (A^*)^2$,

$$\operatorname{tr}(A - A^*)(A - A^*)^* = 0.$$

PROOF 3. Use eigenvalues. Let $B = i(A - A^*)$. Then B is Hermitian. We show that B has only zero eigenvalues; consequently, $B = 0$.

Suppose λ is a nonzero eigenvalue of B with eigenvector x:

$$Bx = \lambda x, \quad \lambda \neq 0, \ x \neq 0.$$

Note that the condition $A^*A = A^2$ implies $BA = 0$. We have

$$\lambda A^*x = A^*(Bx) = (BA)^*x = 0.$$

Thus, $A^*x = 0$. But $Bx = \lambda x$ yields $A^*x = Ax + i\lambda x$. Therefore,

$$0 = x^*A^*x = x^*Ax + i\lambda x^*x = \overline{x^*A^*x} + i\lambda x^*x = i\lambda x^*x.$$

It follows that $\lambda = 0$, a contradiction to the assumption $\lambda \neq 0$.

PROOF 4. Use inner product. Note that (Problem 15, Section 1.4)

$$\mathbb{C}^n = \operatorname{Ker} A^* \oplus \operatorname{Im} A.$$

Thus, to show $A^* = A$, it suffices to show $A^*x = Ax$ for every $x \in \operatorname{Ker} A^*$ and $x \in \operatorname{Im} A$. If $x \in \operatorname{Ker} A^*$, then, by $A^2 = (A^*)^2 = A^*A$,

$$(Ax, Ax) = (A^*Ax, x) = ((A^*)^2 x, x) = 0.$$

This forces $Ax = 0$, namely, $Ax = A^*x$ for every $x \in \operatorname{Ker} A^*$. If $x \in \operatorname{Im} A$, write $x = Ay$, $y \in \mathbb{C}^n$. We then have

$$A^*x = (A^*A)y = A^2y = A(Ay) = Ax. \ \blacksquare$$

Problems

1. What are the differences in decompositions between normal, Hermitian, positive semidefinite, and unitary matrices?

2. Show that the diagonal entries of a Hermitian matrix are all real.

3. Show that $A^* + A$ and A^*A are Hermitian for any square matrix A.

4. Show that if A and B are Hermitian matrices of the same size, then so are $A + B$ and $A - B$. What about AB?

5. Let A be an n-square Hermitian matrix. Show that C^*AC is also Hermitian for any $n \times m$ complex matrix C.

6. Show that if A and B are Hermitian matrices of the same size, then $AB = 0 \Leftrightarrow BA = 0$. What if A and B are not Hermitian?

7. Is a matrix similar to a Hermitian matrix necessarily Hermitian? What if unitary similarity is assumed?

8. Show that if matrix A is *skew-Hermitian*, that is, $A^* = -A$, then $A = iB$ for some Hermitian matrix B.

9. Show that a square complex matrix A can be uniquely written as
$$A = B + iC = S - iT,$$
where B and C are Hermitian, S and T are skew-Hermitian.

10. Show directly the implication (4)$\Rightarrow$(1) in Theorem 7.2.

11. If A is Hermitian, show that A^2 is positive semidefinite.

12. Find a unitary matrix U such that U^*HU is diagonal, where
$$H = \begin{pmatrix} 1 & -i \\ i & 1 \end{pmatrix}.$$

13. Let A and B be Hermitian matrices of the same size. If $AB - BA$ and $A - B$ commute, show that A and B commute.

14. Let $A \in M_n$. Show that A is Hermitian if and only if
$$(Ax, y) = (x, Ay), \quad x, y \in \mathbb{C}^n,$$
and if and only if
$$(Ax, x) = (x, Ax), \quad x \in \mathbb{C}^n.$$
Is it true that A is Hermitian if
$$(Ax, x) = (x, Ax), \quad x \in \mathbb{R}^n?$$

15. Let A be an n-square complex matrix. Show that

    (a) $\operatorname{tr}(AX) = 0$ for all Hermitian $X \in \mathbb{M}_n$ if and only if $A = 0$,
    (b) $\operatorname{tr}(AX) \in \mathbb{R}$ for all Hermitian $X \in \mathbb{M}_n$ if and only if $A = A^*$,
    (c) if A is Hermitian and $\operatorname{tr} A \geq \operatorname{Re}\operatorname{tr}(AU)$ for all unitary $U \in \mathbb{M}_n$, then $A \geq 0$.

16. Show that the rank of a Hermitian matrix is the same as the number of nonzero eigenvalues of the matrix and that the rank of a general matrix A equals the number of nonzero singular values of A, but not the number of nonzero eigenvalues (in general).

17. Show that if A is a Hermitian matrix of rank r, then A has a nonsingular principal submatrix of order r. How about a general matrix?

18. Let A be a Hermitian matrix with rank r. Show that all nonzero $r \times r$ principal minors of A have the same sign.

19. Let $A = A^*$ and $\operatorname{tr} A = 0$. If the sum of all 2×2 principal minors of A is zero, show that $A = 0$. (Hint: Use Problem 14, Section 4.4.)

20. Show that the numbers of positive, negative, and zero eigenvalues in Theorem 7.1 do not depend on the choice of unitary matrix U.

21. Let A be an $n \times n$ Hermitian matrix. The *inertia* of A is defined to be the ordered triple (i_+, i_-, i_0), where i_+, i_-, and i_0 are the numbers of positive, negative, and zero eigenvalues of A, respectively. Denote the inertia of A by $\operatorname{In}(A)$. Show that for any invertible $P \in \mathbb{M}_n$

$$\operatorname{In}(P^*AP) = \operatorname{In}(A)$$

and for any $n \times m$ matrix Q with rank r

$$i_-(Q^*AQ) \leq r, \quad i_+(Q^*AQ) \leq r.$$

22. Let M be a Hermitian matrix. If A is a nonsingular principal submatrix of M and $\widetilde{A}$ is the Schur complement of A in M, show that

$$\operatorname{In}(M) = \operatorname{In}(A) + \operatorname{In}(\widetilde{A}).$$

23. Show that for any $m \times n$ complex matrix A,

$$\operatorname{In}\begin{pmatrix} I_m & A \\ A^* & I_n \end{pmatrix} = (m, 0, 0) + \operatorname{In}(I_n - A^*A) = (n, 0, 0) + \operatorname{In}(I_m - AA^*).$$

7.2 The Product of Hermitian Matrices

This section concerns the product of two Hermitian matrices. As is known, the product of two Hermitian matrices is not necessarily Hermitian in general. For instance, take

$$A = \begin{pmatrix} 1 & 0 \\ 0 & -2 \end{pmatrix}, \quad B = \begin{pmatrix} 1 & 1 \\ 1 & -1 \end{pmatrix}.$$

Note that the eigenvalues of AB are the nonreal numbers $\frac{1}{2}(3 \pm \sqrt{7}\,i)$.

We first show a trace inequality of the product of two Hermitian matrices, and then we turn our attention to discussing when a product is Hermitian. Note that the trace of a product of two Hermitian matrices is always real although the product is not Hermitian. This is seen as follows: If A and B are Hermitian, then

$$\operatorname{tr}(AB) = \operatorname{tr}(A^*B^*) = \operatorname{tr}(BA)^* = \overline{\operatorname{tr}(BA)} = \overline{\operatorname{tr}(AB)}.$$

That is, $\operatorname{tr}(AB)$ is real. Is this true for three Hermitian matrices?

Theorem 7.3 *Let A and B be n-square Hermitian matrices. Then*

$$\operatorname{tr}(AB)^2 \leq \operatorname{tr}(A^2B^2). \tag{7.2}$$

Equality occurs if and only if A and B commute, namely, $AB = BA$.

PROOF 1. Let $C = AB - BA$. Using the fact that $\operatorname{tr}(XY) = \operatorname{tr}(YX)$ for any square matrices X and Y of the same size, we compute

$$\begin{aligned}\operatorname{tr}(C^*C) &= \operatorname{tr}(BA - AB)(AB - BA) \\ &= \operatorname{tr}(BA^2B) + \operatorname{tr}(AB^2A) - \operatorname{tr}(BABA) - \operatorname{tr}(ABAB) \\ &= 2\operatorname{tr}(A^2B^2) - 2\operatorname{tr}(AB)^2.\end{aligned}$$

Note that $\operatorname{tr}(A^2B^2) = \operatorname{tr}(AB^2A)$ is real since AB^2A is Hermitian. Thus $\operatorname{tr}(AB)^2$ is real. The inequality (7.2) then follows immediately from the fact that $\operatorname{tr}(C^*C) \geq 0$ with equality if and only if $C = 0$, that is, $AB = BA$.

PROOF 2. Since for any unitary matrix $U \in \mathbb{M}_n$, U^*AU is also Hermitian, inequality (7.2) holds if and only if

$$\operatorname{tr}\left((U^*AU)B\right)^2 \leq \operatorname{tr}\left((U^*AU)^2 B^2\right).$$

Thus we assume $A = \operatorname{diag}(a_1, \ldots, a_n)$ by Schur decomposition. Then

$$\begin{aligned}\operatorname{tr}(A^2 B^2) - \operatorname{tr}(AB)^2 &= \sum_{i,j} a_i^2 |b_{ij}|^2 - \sum_{i,j} a_i a_j |b_{ij}|^2 \\ &= \sum_{i<j} (a_i - a_j)^2 |b_{ij}|^2 \geq 0\end{aligned}$$

with equality if and only if $a_i b_{ij} = a_j b_{ij}$. This implies $AB = BA$.

PROOF 3 FOR THE EQUALITY CASE. We use the fact that a matrix X is Hermitian if and only if $\operatorname{tr} X^2 = \operatorname{tr}(XX^*)$, by Theorem 7.2(4).

Notice that the hermitity of A and B gives

$$\operatorname{tr}(A^2 B^2) = \operatorname{tr}(ABBA) = \operatorname{tr}(AB)(AB)^*.$$

Thus,

$$\operatorname{tr}(AB)^2 = \operatorname{tr}(A^2 B^2) \;\;\Rightarrow\;\; \operatorname{tr}(AB)^2 = \operatorname{tr}(AB)(AB)^*.$$

It follows that AB is Hermitian. Hence,

$$AB = (AB)^* = B^* A^* = BA. \;\blacksquare$$

Clearly, the product of two Hermitian matrices is Hermitian if and only if these two matrices commute (Problem 4). We are now interested in the following question: When can a given matrix be written as a product of two Hermitian matrices?

Let A be given. Suppose $A = BC$ is a product of two Hermitian matrices B and C of the same size. If B is nonsingular, then

$$A = BC = B(CB)B^{-1} = BA^* B^{-1},$$

namely, A is similar to A^*. This is in fact a necessary and sufficient condition for a matrix to be a product of two Hermitian matrices.

Theorem 7.4 *A square matrix A is a product of two Hermitian matrices if and only if A is similar to A^*.*

PROOF. Necessity: Let $A = BC$, where B and C are n-square Hermitian matrices. Then we have at once

$$AB = BCB = B(BC)^* = BA^*$$

and inductively for every positive integer k

$$A^k B = B(A^*)^k. \tag{7.3}$$

We may write, without loss of generality via similarity (Problem 7),

$$A = \begin{pmatrix} J & 0 \\ 0 & K \end{pmatrix},$$

where J and K contain the Jordan blocks of eigenvalues 0 and nonzero, respectively. Note that J is nilpotent and K is invertible.

Partition B and C conformally with A as

$$B = \begin{pmatrix} L & M \\ M^* & N \end{pmatrix}, \quad C = \begin{pmatrix} P & Q \\ Q^* & R \end{pmatrix}.$$

Then (7.3) implies that for each positive integer k

$$K^k M^* = M^* (J^*)^k.$$

Notice that $(J^*)^k = 0$ when $k \geq n$, for J is nilpotent. It follows that $M = 0$, since K is nonsingular. Thus $A = BC$ is the same as

$$\begin{pmatrix} J & 0 \\ 0 & K \end{pmatrix} = \begin{pmatrix} L & 0 \\ 0 & N \end{pmatrix} \begin{pmatrix} P & Q \\ Q^* & R \end{pmatrix}.$$

This yields $K = NR$, and hence N and R are nonsingular.

Taking $k = 1$ in (7.3), we have

$$\begin{pmatrix} J & 0 \\ 0 & K \end{pmatrix} \begin{pmatrix} L & 0 \\ 0 & N \end{pmatrix} = \begin{pmatrix} L & 0 \\ 0 & N \end{pmatrix} \begin{pmatrix} J^* & 0 \\ 0 & K^* \end{pmatrix},$$

which gives $KN = NK^*$, or, since N is invertible,

$$N^{-1} K N = K^*.$$

In other words, K is similar to K^*. On the other hand, any square matrix is similar to its transpose (Theorem 3.13(1)). Thus J is similar to $J^T = J^*$, and it follows that A is similar to A^*.

Sufficiency: We show that if A is similar to A^*, then A can be expressed as a product of two Hermitian matrices. Notice that

$$A = P^{-1}H_1H_2P \quad \Rightarrow \quad A = P^{-1}H_1(P^{-1})^*P^*H_2P.$$

This says if A is similar to a product of Hermitian matrices, then A is in fact a product of Hermitian matrices.

Recall from Theorem 3.13(2) that A is similar to A^* if and only if the Jordan blocks of the nonreal eigenvalues λ of A occur in conjugate pairs. Thus it is sufficient to show that the paired Jordan block

$$\begin{pmatrix} J(\lambda) & 0 \\ 0 & J(\bar{\lambda}) \end{pmatrix},$$

where $J(\lambda)$ is a Jordan block with λ on the diagonal, is similar to a product of two Hermitian matrices. This is seen as follows: Matrices

$$\begin{pmatrix} J(\lambda) & 0 \\ 0 & J(\bar{\lambda}) \end{pmatrix} \quad \text{and} \quad \begin{pmatrix} J(\lambda) & 0 \\ 0 & (J(\bar{\lambda}))^T \end{pmatrix}$$

are similar, since any square matrix is similar to its transpose. But

$$\begin{pmatrix} J(\lambda) & 0 \\ 0 & (J(\bar{\lambda}))^T \end{pmatrix} = \begin{pmatrix} J(\lambda) & 0 \\ 0 & (J(\lambda))^* \end{pmatrix},$$

which is equal to a product of two Hermitian matrices:

$$\begin{pmatrix} 0 & J(\lambda) \\ (J(\lambda))^* & 0 \end{pmatrix} \begin{pmatrix} 0 & I \\ I & 0 \end{pmatrix}. \blacksquare$$

Problems _____

1. Let A and B be Hermitian matrices of the same size. Show that $AB - BA$ is skew-Hermitian and $ABA - BAB$ is Hermitian.

2. Let A, B, and C be $n \times n$ Hermitian matrices. Prove or disprove

$$\text{tr}(ABC) = \text{tr}(BCA) \quad \text{or} \quad \text{tr}(ABC) = \text{tr}(CBA).$$

Is $\text{tr}(ABC)$ necessarily real? How about $\det(ABC)$? Eigenvalues?

3. Let A and B be $n \times n$ Hermitian matrices. Show that $\operatorname{tr}(A^k B^k)$ and $\operatorname{tr}(AB)^k$ are real for any positive integer k.

4. Let A and B be n-square Hermitian matrices. Show that the product AB is Hermitian if and only if $AB = BA$.

5. Let A and B be Hermitian matrices of the same size. Show that AB and BA are similar.

6. If λ are the eigenvalues of a Hermitian matrix A, what are the singular eigenvalues and the singular value decomposition of A?

7. Give in detail the reason why the matrix A may be assumed to be a Jordan form in the proof of Theorem 7.4.

8. Let $A \in \mathbb{M}_n$. If $A^k = A^{k+1}$ for some positive integer k, show that
$$\operatorname{tr} A = \operatorname{tr} A^2 = \cdots = \operatorname{tr} A^n = \cdots.$$

9. Show that for any square complex matrices A and B of the same size
$$\operatorname{tr}(AB - BA) = 0, \quad \operatorname{tr}(AB - BA)(AB + BA) = 0.$$

10. Let A and B be Hermitian matrices of the same size. Show that
   (a) $|\operatorname{tr}(AB)| \leq \operatorname{tr}(\frac{A+B}{2})^2$,
   (b) $|\operatorname{tr}(AB)| \leq (\operatorname{tr} A^2)^{\frac{1}{2}} (\operatorname{tr} B^2)^{\frac{1}{2}}$,
   (c) $(\operatorname{tr}(A+B)^2)^{\frac{1}{2}} \leq (\operatorname{tr} A^2)^{\frac{1}{2}} + (\operatorname{tr} B^2)^{\frac{1}{2}}$.

11. Let A, B, and C be Hermitian matrices of the same size. Show that
$$|\operatorname{tr}(ABC)| \leq (\operatorname{tr} A^2 B^2 C^2)^{\frac{1}{2}}$$
is not true in general. (Hint: Assume that A is a diagonal matrix.)

12. Let A be a square matrix with all eigenvalues real (A is not necessarily Hermitian), k of which are nonzero, $k \geq 1$. Show that
$$\frac{(\operatorname{tr} A)^2}{\operatorname{tr} A^2} \leq k.$$

13. Let $A, B \in \mathbb{M}_n$ be Hermitian matrices of positive traces. Show that
$$\frac{\operatorname{tr}(A+B)^2}{\operatorname{tr}(A+B)} \leq \frac{\operatorname{tr} A^2}{\operatorname{tr} A} + \frac{\operatorname{tr} B^2}{\operatorname{tr} B}.$$

14. Let A and B be Hermitian matrices of the same size. Show that there exists a unitary matrix U such that U^*AU and U^*BU are both diagonal if and only if $AB = BA$.

15. Let A and B be Hermitian matrices of the same size. If $AB = BA$, show that for any $a, b \in \mathbb{C}$ the eigenvalues of $aA + bB$ are in the form $a\lambda + b\mu$, where λ and μ are some eigenvalues of A and B, respectively.

16. Let $A \in \mathbb{M}_n$ and let S be an invertible matrix with $S^{-1}AS = A^*$. Set

$$H_c = cS + \bar{c}S^*.$$

Show that H_c and AH_c are Hermitian matrices. Also show that

$$A = (AH_c)H_c^{-1}$$

is a product of two Hermitian matrices for some c such that H_c is invertible. Why does such an invertible matrix H_c exist?

17. Show that a matrix is diagonalizable (not necessarily unitarily diagonalizable) with real eigenvalues if and only if it can be written as a product of a positive definite matrix and a Hermitian matrix.

18. Show that any square matrix is a product of two symmetric matrices.

19. Let $A \in \mathbb{M}_n$ be positive definite and let $B \in \mathbb{M}_n$ be Hermitian such that AB is a Hermitian matrix. Show that AB is positive definite if and only if the eigenvalues of B are all positive.

20. Let $A \in \mathbb{M}_n$ be a Hermitian matrix. Show that $\operatorname{tr} A > 0$ if and only if $A = B + B^*$ for some B similar to a positive definite matrix.

21. Show that a matrix A is a product of two positive semidefinite matrices if and only if A is similar to a positive semidefinite matrix.

22. Show that the product AB of a positive definite matrix A and a Hermitian matrix B is diagonalizable. What if A is singular?

23. Let A be an n-square complex matrix. If $A + A^* > 0$, show that every eigenvalue of A has a positive real part. Use this fact to show that $X > 0$ if X is a Hermitian matrix satisfying for some $Y > 0$

$$XY + YX > 0.$$

7.3 The Min-Max Theorem and Interlacing Theorem

In this section we make use of some techniques on vector spaces to derive eigenvalue inequalities for Hermitian matrices. The idea is to choose vectors in certain subspaces spanned by eigenvectors in order to obtain the min-max representations and the desired inequalities.

Let H be an $n \times n$ Hermitian matrix with (necessarily real) eigenvalues $\lambda_i(H)$, or simply λ_i, $i = 1, 2, \ldots, n$. By Theorem 7.1, there is a unitary matrix U such that

$$U^* H U = \mathrm{diag}(\lambda_1, \lambda_2, \ldots, \lambda_n)$$

or

$$HU = U \, \mathrm{diag}(\lambda_1, \lambda_2, \ldots, \lambda_n).$$

The column vectors $u_1, u_2, \ldots, u_n$ of U are orthonormal eigenvectors of H corresponding to $\lambda_1, \lambda_2, \ldots, \lambda_n$, respectively, that is,

$$Hu_i = \lambda_i u_i, \quad u_i^* u_j = \delta_{ij}, \quad i, j = 1, 2, \ldots, n, \tag{7.4}$$

where $\delta_{ij} = 1$ if $i = j$, and 0 otherwise.

We assume that the eigenvalues and singular values of a Hermitian matrix H are arranged in decreasing order:

$$\lambda_{\max} = \lambda_1 \geq \lambda_2 \geq \cdots \geq \lambda_n = \lambda_{\min};$$

$$\sigma_{\max} = \sigma_1 \geq \sigma_2 \geq \cdots \geq \sigma_n = \sigma_{\min}.$$

The following theorem is of fundamental importance to the rest of this chapter. The idea and result will be employed frequently.

Theorem 7.5 *Let $H \in \mathbb{M}_n$ be a Hermitian matrix. Let u_i be orthonormal eigenvectors of H corresponding to the eigenvalues λ_i, $i = 1, 2, \ldots, n$, and let $W = \mathrm{Span}\{u_p, \ldots, u_q\}$, where $1 \leq p \leq q \leq n$. Then for any unit $x \in W$*

$$\lambda_q(H) \leq x^* H x \leq \lambda_p(H). \tag{7.5}$$

PROOF. Let $x = x_p u_p + \cdots + x_q u_q$. Then by using (7.4)

$$\begin{aligned} x^* H x &= x^*(x_p H u_p + \cdots + x_q H u_q) \\ &= x^*(\lambda_p x_p u_p + \cdots + \lambda_q x_q u_q) \\ &= \lambda_p x_p x^* u_p + \cdots + \lambda_q x_q x^* u_q \\ &= \lambda_p |x_p|^2 + \cdots + \lambda_q |x_q|^2. \end{aligned}$$

The inequality follows since x is a unit vector: $\sum_{i=p}^{q} |x_i|^2 = 1$. ∎

Theorem 7.6 (Rayleigh-Ritz) *Let $H \in \mathbb{M}_n$ be Hermitian. Then*

$$\lambda_{\min}(H) = \min_{x^* x = 1} x^* H x$$

and

$$\lambda_{\max}(H) = \max_{x^* x = 1} x^* H x.$$

PROOF. The eigenvectors of H in (7.4) form an orthonormal basis for $\mathbb{C}^n$. By (7.5), it is sufficient to observe that

$$\lambda_{\min}(H) = u_n^* H u_n \quad \text{and} \quad \lambda_{\max}(H) = u_1^* H u_1. \ \blacksquare$$

Recall the dimension identity (Theorem 1.1): If S_1 and S_2 are subspaces of an n-dimensional vector space, then

$$\dim(S_1 \cap S_2) = \dim S_1 + \dim S_2 - \dim(S_1 + S_2).$$

It follows that $S_1 \cap S_2$ is nonempty if

$$\dim S_1 + \dim S_2 > n \tag{7.6}$$

and that for three subspaces S_1, S_2, and S_3,

$$\dim(S_1 \cap S_2 \cap S_3) \geq \dim S_1 + \dim S_2 + \dim S_3 - 2n. \tag{7.7}$$

We will use these inequalities to obtain the min-max theorem and derive eigenvalue inequalities for Hermitian matrices.

Theorem 7.7 (Courant-Fischer) *Let $H \in \mathbb{M}_n$ be Hermitian. Then*

$$\lambda_k(H) = \max_{S^k} \min_{x \in S^k,\, x^* x = 1} x^* H x,$$

where S^k denotes an arbitrary k-dimensional subspace of $\mathbb{C}^n$.

SEC. 7.3 THE MIN-MAX THEOREM AND INTERLACING THEOREM

PROOF. Let
$$S_1 = \text{Span}\{u_k, \ldots, u_n\}, \quad \dim S_1 = n - k + 1,$$
and let
$$S_2 = S^k, \quad \text{any } k\text{-dimensional subspace of } \mathbb{C}^n.$$
By (7.6), there exists a vector x such that
$$x \in S_1 \cap S_2, \quad x^*x = 1,$$
and for this x, by (7.5),
$$\lambda_k \geq x^* H x.$$
Thus, for any k-dimensional space S^k,
$$\lambda_k \geq \min_{v \in S^k,\, v^*v=1} v^* H v.$$
It follows that
$$\lambda_k \geq \max_{S^k} \min_{v \in S^k,\, v^*v=1} v^* H v.$$
However, for any unit vector $v \in \text{Span}\{u_1, u_2, \ldots, u_k\}$, a k-dimensional subspace, we have by (7.5) again
$$v^* H v \geq \lambda_k \quad \text{and} \quad u_k^* H u_k = \lambda_k.$$
Thus, for $S^k = \text{Span}\{u_1, \ldots, u_k\}$,
$$\min_{v \in S^k,\, v^*v=1} v^* H v \geq \lambda_k.$$
It follows that
$$\max_{S^k} \min_{v \in S^k,\, v^*v=1} v^* H v \geq \lambda_k.$$
Putting these together,
$$\max_{S^k} \min_{v \in S^k,\, v^*v=1} v^* H v = \lambda_k. \quad \blacksquare$$

The following theorem is usually referred to as the eigenvalue *interlacing theorem*, also known as the *Cauchy, Poincaré,* or *Sturm interlacing theorem.* It states, simply put, that the eigenvalues of a principal submatrix of a Hermitian matrix interlace those of the underlying matrix. This is used to obtain many matrix inequalities.

Theorem 7.8 (Interlacing Theorem) *Let H be an $n \times n$ Hermitian matrix partitioned as*

$$H = \begin{pmatrix} A & B \\ B^* & C \end{pmatrix},$$

where A is an $m \times m$, $1 \leq m \leq n$, principal submatrix of H. Then

$$\lambda_{k+n-m}(H) \leq \lambda_k(A) \leq \lambda_k(H), \quad k = 1, 2, \ldots, m.$$

In particular, when $m = n - 1$,

$$\lambda_n(H) \leq \lambda_{n-1}(A) \leq \lambda_{n-1}(H) \leq \cdots \leq \lambda_2(H) \leq \lambda_1(A) \leq \lambda_1(H).$$

We present three different proofs in the following. Denote for convenience the eigenvalues of H and A, respectively, by

$$\lambda_1 \geq \lambda_2 \geq \cdots \geq \lambda_n, \quad \mu_1 \geq \mu_2 \geq \cdots \geq \mu_m.$$

PROOF 1. Use subspaces spanned by certain eigenvectors. Let u_i and v_i be orthonormal eigenvectors of H and A belonging to the eigenvalues λ_i and μ_i, respectively. Symbolically,

$$Hu_i = \lambda_i u_i, \quad u_i^* u_j = \delta_{ij}, \quad i, j = 1, 2, \ldots, n, \quad u_i \in \mathbb{C}^n,$$

$$Av_i = \mu_i v_i, \quad v_i^* v_j = \delta_{ij}, \quad i, j = 1, 2, \ldots, m, \quad v_i \in \mathbb{C}^m.$$

Let

$$w_i = \begin{pmatrix} v_i \\ 0 \end{pmatrix} \in \mathbb{C}^n, \quad i = 1, 2, \ldots, m.$$

Note that the w_i are eigenvectors belonging to the respective eigenvalues μ_i of the partitioned matrix $A \oplus 0$. For $1 \leq k \leq m$, we set

$$S_1 = \text{Span}\{u_k, \ldots, u_n\}$$

and

$$S_2 = \text{Span}\{w_1, \ldots, w_k\}.$$

Then

$$\dim S_1 = n - k + 1 \quad \text{and} \quad \dim S_2 = k.$$

Sec. 7.3   The Min-Max Theorem and Interlacing Theorem

We thus have a vector x such that
$$x \in S_1 \cap S_2, \quad x^*x = 1,$$
and for this x, by (7.5),
$$\lambda_k \geq x^*Hx \geq \mu_k. \tag{7.8}$$

An application of this inequality to $-H$ gives $\mu_k \geq \lambda_{k+n-m}$.

PROOF 2. Use the adjoint matrix and continuity of functions. Reduce the proof to the case $m = n - 1$ by considering a sequence of leading principal submatrices, two consecutive ones differing in size by one.

We may assume that $\lambda_1 > \lambda_2 > \cdots > \lambda_n$. The case in which some of the λ_i are equal will follow from a continuity argument (replacing λ_i with $\lambda_i + \epsilon_i$). Let U be an n-square unitary matrix such that
$$H = U^* \operatorname{diag}(\lambda_1, \lambda_2, \ldots, \lambda_n) U.$$
Then
$$tI - H = U^* \operatorname{diag}(t - \lambda_1, t - \lambda_2, \ldots, t - \lambda_n) U \tag{7.9}$$
and for $t \neq \lambda_i$, $i = 1, 2, \ldots, n$,
$$\operatorname{adj}(tI - H) = \det(tI - H)(tI - H)^{-1}. \tag{7.10}$$
Upon computation, the (n,n)-entry of $(tI - H)^{-1}$ by using (7.9) is
$$\frac{|u_{1n}|^2}{t - \lambda_1} + \frac{|u_{2n}|^2}{t - \lambda_2} + \cdots + \frac{|u_{nn}|^2}{t - \lambda_n}$$
and the (n,n)-entry of $\operatorname{adj}(tI - H)$ is $\det(tI - A)$. Thus by (7.10)
$$\frac{\det(tI - A)}{\det(tI - H)} = \frac{|u_{1n}|^2}{t - \lambda_1} + \frac{|u_{2n}|^2}{t - \lambda_2} + \cdots + \frac{|u_{nn}|^2}{t - \lambda_n}. \tag{7.11}$$

Notice that the function of t defined in (7.11) is continuous except at the points λ_i, and that it is decreasing on each interval $(\lambda_{i+1}, \lambda_i)$. On the other hand, since $\mu_1, \mu_2, \ldots, \mu_n$ are the roots of the numerator $\det(tI - A)$, it follows, by considering the behavior of the function over the intervals divided by the eigenvalues λ_i, that
$$\mu_i \in [\lambda_{i+1}, \lambda_i], \quad i = 1, 2, \ldots, n-1.$$

The preceding proof is surely a good example of applications of calculus to linear algebra and matrix theory.

PROOF 3. Use the Courant-Fischer theorem. Let $1 \leq k \leq m$. Then

$$\lambda_k(A) = \max_{S_m^k} \min_{x \in S_m^k,\ x^*x=1} x^*Ax,$$

where S_m^k is an arbitrary k-dimensional subspace of $\mathbb{C}^m$, and

$$\lambda_k(H) = \max_{S_n^k} \min_{x \in S_n^k,\ x^*x=1} x^*Hx,$$

where S_n^k is an arbitrary k-dimensional subspace of $\mathbb{C}^n$.

Denote by S_0^k the k-dimensional subspace of $\mathbb{C}^n$ of the vectors

$$y = \begin{pmatrix} x \\ 0 \end{pmatrix}, \quad \text{where } x \in S_m^k.$$

Noticing that $y^*Hy = x^*Ax$, we have, by a simple computation,

$$\begin{aligned}
\lambda_k(H) &= \max_{S_n^k} \min_{x \in S_n^k,\ x^*x=1} x^*Hx \\
&\geq \max_{S_0^k} \min_{y \in S_0^k,\ y^*y=1} y^*Hy \\
&= \max_{S_m^k} \min_{x \in S_m^k,\ x^*x=1} x^*Ax \\
&= \lambda_k(A).
\end{aligned}$$

The other inequality is obtained by replacing H with $-H$. ∎

As an application of the interlacing theorem, we present a result due to Poincaré: If $A \in \mathbb{M}_m$ is a Hermitian matrix, then for any $m \times n$ matrix V satisfying $V^*V = I_n$ and for each $i = 1, 2, \ldots, m$,

$$\lambda_{i+m-n}(A) \leq \lambda_i(V^*AV) \leq \lambda_i(A). \tag{7.12}$$

To see this, first notice that $V^*V = I_n \Rightarrow m \geq n$ (Problem 1). Let U be an $m \times (m-n)$ matrix such that (V, U) is unitary. Then

$$(V, U)^* A (V, U) = \begin{pmatrix} V^*AV & V^*AU \\ U^*AV & U^*AU \end{pmatrix}.$$

Thus by applying the interlacing theorem, we have

$$\begin{aligned}\lambda_{i+m-n}(A) &= \lambda_{i+m-n}\big((V,U)^*A(V,U)\big)\\ &\leq \lambda_i(V^*AV)\\ &\leq \lambda_i\big((V,U)^*A(V,U)\big)\\ &= \lambda_i(A).\end{aligned}$$

Problems

1. Let V be an $m \times n$ matrix. Show that if $V^*V = I_n$, then $m \geq n$.

2. Let $[A]$ be a principal submatrix of A. If A is Hermitian, show that
$$\lambda_{\min}(A) \leq \lambda_{\min}([A]) \leq \lambda_{\max}([A]) \leq \lambda_{\max}(A).$$

3. Let $\lambda_k(A)$ denote the kth largest eigenvalue of an n-square Hermitian matrix A. Show that if A is nonsingular, then
$$\lambda_k(A^{-1}) = \frac{1}{\lambda_{n-k+1}(A)}.$$

4. Let $A \in \mathbb{M}_n$ be Hermitian. Show that for every nonzero $x \in \mathbb{C}^n$
$$\lambda_{\min}(A) \leq \frac{x^*Ax}{x^*x} \leq \lambda_{\max}(A)$$
and for all diagonal entries a_{ii} of A
$$\lambda_{\min}(A) \leq a_{ii} \leq \lambda_{\max}(A).$$

5. For any Hermitian matrices A and B of the same size, show that
$$\lambda_{\max}(A-B) + \lambda_{\min}(B) \leq \lambda_{\max}(A).$$

6. Let $A \in \mathbb{M}_n$ be Hermitian and $B \in \mathbb{M}_n$ be positive definite. Show that the eigenvalues of AB^{-1} are all real, that
$$\lambda_{\max}(AB^{-1}) = \max_{x \neq 0} \frac{x^*Ax}{x^*Bx},$$
and that
$$\lambda_{\min}(AB^{-1}) = \min_{x \neq 0} \frac{x^*Ax}{x^*Bx}.$$

7. Let $\{x_1, \ldots, x_k\}$ and $\{y_1, \ldots, y_k\}$ be two sets of orthonormal vectors in $\mathbb{C}^n$. If $\text{Span}\{x_1, \ldots, x_k\} = \text{Span}\{y_1, \ldots, y_k\}$, show that
$$\sum_{i=1}^{k} x_i^* A x_i = \sum_{i=1}^{k} y_i^* A y_i, \quad \text{for every } A \in \mathbb{M}_n.$$

8. Let A be a Hermitian matrix with the largest and smallest eigenvalues in absolute value $|\lambda|_{\max}$ and $|\lambda|_{\min}$, respectively. Show that
$$|\lambda|_{\max} = \max_{x^* x = 1} |x^* A x|$$
but
$$|\lambda|_{\min} = \min_{x^* x = 1} |x^* A x|$$
is not true in general.

9. Let A be an $m \times n$ complex matrix with the ith largest singular value σ_i. Denote by S^k a k-dimensional subspace of $\mathbb{C}^n$. Show that
$$\sigma_i = \max_{S^k} \min_{x \in S^k, x^* x = 1} x^* (A^* A)^{\frac{1}{2}} x = \max_{S^k} \min_{x \in S^k, x^* x = 1} (x^* A^* A x)^{\frac{1}{2}}.$$

10. Let A be an n-square positive semidefinite matrix. If V is an $n \times m$ complex matrix such that $V^* V = I_m$, show that
$$\prod_{i=1}^{m} \lambda_{n-m+i}(A) \leq \det(V^* A V) \leq \prod_{i=1}^{m} \lambda_i(A).$$

11. Let A be a positive semidefinite matrix partitioned as
$$A = \begin{pmatrix} A_{11} & A_{12} \\ A_{21} & A_{22} \end{pmatrix},$$
where A_{11} is square, and let $\widetilde{A_{11}} = A_{22} - A_{21} A_{11}^{-1} A_{12}$ be the Schur complement of A_{11} in A when A_{11} is nonsingular. Show that
$$\lambda_{\min}(A) \leq \lambda_{\min}(\widetilde{A_{11}}) \leq \lambda_{\min}(A_{22}).$$

12. Let A be an $n \times n$ Hermitian matrix with the largest eigenvalue $\lambda_{\max}$ and the smallest eigenvalue $\lambda_{\min}$. Show that
$$\lambda_{\max} - \lambda_{\min} = 2 \max\{|x^* A y| : x, y \in \mathbb{C}^n \text{ orthonormal}\}.$$
Derive that
$$\lambda_{\max} - \lambda_{\min} \geq 2 \max_{i,j} |a_{ij}|.$$

7.4 Eigenvalue and Singular Value Inequalities

The goal of this section is to derive some basic eigenvalue and singular value inequalities by using the min-max theorem and the interlacing theorem. We assume that the eigenvalues, singular values, and diagonal entries of Hermitian matrices are arranged in decreasing order.

The following theorem on comparing two Hermitian matrices best characterizes the Löwner ordering in terms of eigenvalues.

Theorem 7.9 *Let $A, B \in \mathbb{M}_n$ be Hermitian matrices. Then*

$$A \geq B \quad \Rightarrow \quad \lambda_i(A) \geq \lambda_i(B), \quad i = 1, 2, \ldots, n.$$

This follows from the Fischer-Courant theorem immediately, for

$$A \geq B \quad \Rightarrow \quad x^*Ax \geq x^*Bx, \quad x \in \mathbb{C}^n.$$

Our next theorem compares the eigenvalues of sum, ordinary and Hadamard products of matrices to those of the individual matrices.

Theorem 7.10 *Let $A, B \in \mathbb{M}_n$ be Hermitian matrices. Then*

1. $\lambda_i(A) + \lambda_n(B) \leq \lambda_i(A + B) \leq \lambda_i(A) + \lambda_1(B),$
2. $\lambda_i(A)\lambda_n(B) \leq \lambda_i(AB) \leq \lambda_i(A)\lambda_1(B)$ *if $A \geq 0$ and $B \geq 0$,*
3. $a_{ii}\lambda_n(B) \leq \lambda_i(A \circ B) \leq a_{ii}\lambda_1(B)$ *if $A \geq 0$ and $B \geq 0$.*

PROOF. Let x be unit vectors in $\mathbb{C}^n$. Then we have

$$x^*Ax + \min_x x^*Bx \leq x^*(A+B)x \leq x^*Ax + \max_x x^*Bx.$$

Thus

$$x^*Ax + \lambda_n(B) \leq x^*(A+B)x \leq x^*Ax + \lambda_1(B).$$

An application of the min-max theorem results in (1). To show (2), we write $\lambda_i(AB) = \lambda_i(B^{\frac{1}{2}}AB^{\frac{1}{2}})$. Notice that $\lambda_1(A)I - A \geq 0$. Thus

$$B^{\frac{1}{2}}AB^{\frac{1}{2}} \leq B^{\frac{1}{2}}AB^{\frac{1}{2}} + B^{\frac{1}{2}}\left(\lambda_1(A)I - A\right)B^{\frac{1}{2}} = \lambda_1(A)B.$$

An application of Theorem 7.9 gives (2). For (3), recall that the Hadamard product of two positive semidefinite matrices is positive semidefinite (Schur theorem, Section 6.5). Since $\lambda_1(B)I - B \geq 0$ and $B - \lambda_n(B)I \geq 0$, we have, by taking the Hadamard product with A,

$$A \circ \left(\lambda_1(B)I - B\right) \geq 0, \quad A \circ \left(B - \lambda_n(B)I\right) \geq 0,$$

which give

$$\lambda_n(B)(I \circ A) \leq A \circ B \leq \lambda_1(B)(I \circ A).$$

Note that $I \circ A = \text{diag}(a_{11}, \ldots, a_{nn})$. Theorem 7.9 gives for each i

$$a_{ii}\lambda_n(B) \leq \lambda_i(A \circ B) \leq a_{ii}\lambda_1(B). \quad \blacksquare$$

It is natural to ask the question: If $A \geq 0$ and $B \geq 0$, is

$$\lambda_i(A)\lambda_n(B) \leq \lambda_i(A \circ B) \leq \lambda_i(A)\lambda_1(B)?$$

The answer is negative (Problem 7). For singular values, we have

Theorem 7.11 *Let A and B be complex matrices. Then*

$$\sigma_i(A) + \lambda_n(B) \leq \sigma_i(A + B) \leq \sigma_i(A) + \sigma_1(B) \tag{7.13}$$

if A and B are of the same size $m \times n$, and

$$\sigma_i(A)\sigma_m(B) \leq \sigma_i(AB) \leq \sigma_i(A)\sigma_1(B) \tag{7.14}$$

if A is $m \times n$ and B is $n \times m$.

PROOF. To show (7.13), notice that the Hermitian matrix

$$\begin{pmatrix} 0 & X \\ X^* & 0 \end{pmatrix},$$

where X is an $m \times n$ matrix with rank r, has eigenvalues

$$\sigma_1(X), \ \ldots, \ \sigma_r(X), \ \overbrace{0, \ldots, 0}^{m+n-2r}, \ -\sigma_r(X), \ \ldots, \ -\sigma_1(X).$$

SEC. 7.4   EIGENVALUE AND SINGULAR VALUE INEQUALITIES   229

Applying the previous theorem to the Hermitian matrices

$$\begin{pmatrix} 0 & A \\ A^* & 0 \end{pmatrix}, \quad \begin{pmatrix} 0 & B \\ B^* & 0 \end{pmatrix}$$

and to their sum, one gets the desired singular value inequalities. For the inequalities on product, it suffices to note that

$$\sigma_i(AB) = \sqrt{\lambda_i(B^*A^*AB)} = \sqrt{\lambda_i(A^*ABB^*)}. \blacksquare$$

Some stronger results can be obtained by using the min-max theorem (Problem 17). For submatrices of a general matrix, we have

Theorem 7.12 *Let B be any submatrix of an $m \times n$ matrix A obtained by deleting a total of r rows and columns. Then*

$$\sigma_{r+i}(A) \leq \sigma_i(B) \leq \sigma_i(A), \quad i = 1, 2, \ldots, \min\{m, n\}.$$

PROOF. We may assume that $r = 1$ and B is obtained from A by deleting a column, say b, that is, $A = (B, b)$. Otherwise one may place B in the upper-left corner of A by permutation and consider a sequence of submatrices of A that contain B, two consecutive ones differing by a row or column (see the second proof of Theorem 7.8).

Notice that B^*B is a principal submatrix of A^*A. Using the eigenvalue interlacing theorem (Theorem 7.8), we have for each i

$$\lambda_{i+1}(A^*A) \leq \lambda_i(B^*B) \leq \lambda_i(A^*A).$$

The proof is completed by taking square roots. $\blacksquare$

We now turn our attention to matrix majorization inequalities. Majorization is a powerful tool in deriving matrix inequalities. (See Marshall and Olkin's book for extensive studies of the topic.)

For two sequences of real numbers arranged in decreasing order,

$$x = (x_1, x_2, \ldots, x_n) \quad \text{and} \quad y = (y_1, y_2, \ldots, y_n),$$

we write $x \prec_w y$ and say x is *weakly majorized* by y if for each $k \leq n$

$$\sum_{i=1}^{k} x_i \leq \sum_{i=1}^{k} y_i,$$

and write $x \prec y$, and say x is *majorized* by y if

$$x \prec_w y \quad \text{and} \quad \sum_{i=1}^n x_i = \sum_{i=1}^n y_i.$$

In addition, we denote the vectors of the eigenvalues, singular values, and the diagonal entries of an n-square Hermitian matrix H, arranged in decreasing order, by, respectively,

$$\lambda(H) = (\lambda_1(H), \lambda_2(H), \ldots, \lambda_n(H)),$$

$$\sigma(H) = (\sigma_1(H), \sigma_2(H), \ldots, \sigma_n(H)),$$

$$d(H) = (d_1(H), d_2(H), \ldots, d_n(H)).$$

Theorem 7.13 (Schur) *Let H be a Hermitian matrix. Then*

$$d(H) \prec \lambda(H).$$

PROOF. Let H_k, $1 \leq k \leq n$, be the principal submatrix of H with the diagonal entries $d_1(H), d_2(H), \ldots, d_k(H)$. Theorem 7.8 yields

$$\sum_{i=1}^k d_i(H) = \operatorname{tr} H_k = \sum_{i=1}^k \lambda_i(H_k) \leq \sum_{i=1}^k \lambda_i(H).$$

Equality holds when $k = n$, for then both sides equal $\operatorname{tr} H$. ∎

It is immediate that if H is a Hermitian matrix and U is any unitary matrix of the same size. Then

$$d(U^*HU) \prec \lambda(H). \tag{7.15}$$

Theorem 7.14 *For any Hermitian matrix $H \in \mathbb{M}_n$ and $1 \leq k \leq n$,*

$$\sum_{i=1}^k \lambda_i(H) = \max_{x_i^* x_j = \delta_{ij}} \sum_{i=1}^k x_i^* H x_i. \tag{7.16}$$

PROOF. Let $U = (V, W)$ be a unitary matrix, where V consists of the orthonormal vectors $x_1, x_2, \ldots, x_k$. Then

$$\sum_{i=1}^{k} x_i^* H x_i = \operatorname{tr}(V^* H V)$$
$$\leq \sum_{i=1}^{k} d_i(U^* H U)$$
$$\leq \sum_{i=1}^{k} \lambda_i(U^* H U) \quad \text{(by (7.15))}$$
$$= \sum_{i=1}^{k} \lambda_i(H).$$

Identity (7.16) follows by choosing the unit eigenvectors x_i of H:

$$\sum_{i=1}^{k} x_i^* H x_i = \sum_{i=1}^{k} \lambda_i(H). \blacksquare$$

Theorem 7.15 (Fan) *Let $A, B \in \mathbb{M}_n$ be Hermitian. Then*

$$\lambda(A + B) \prec \lambda(A) + \lambda(B).$$

This is immediate from an application of the previous theorem to $A + B$. Notice that $\operatorname{tr}(A + B) = \operatorname{tr} A + \operatorname{tr} B$.

Applying Fan's theorem to Hermitian matrices in the form

$$\begin{pmatrix} 0 & X \\ X^* & 0 \end{pmatrix}$$

with replacement of X by A and B, respectively, one gets the following analogous inequalities for singular values.

Theorem 7.16 *Let A and B be $m \times n$ complex matrices. Then*

$$\sigma(A + B) \prec_w \sigma(A) + \sigma(B).$$

As to ordinary and Hadamard products, we have for $A, B \in \mathbb{M}_n$

$$\sigma(AB) \prec_w \sigma(A) \circ \sigma(B)$$

and

$$\sigma(A \circ B) \prec_w \sigma(A) \circ \sigma(B).$$

In particular, for $n \times n$ positive semidefinite matrices A and B,

$$\lambda(AB) \prec_w \lambda(A) \circ \lambda(B)$$

and

$$\lambda(A \circ B) \prec_w \lambda(A) \circ \lambda(B).$$

The proofs of these inequalities are either lengthy or need more advanced tools. We thus omit them here and refer the reader to a paper by R. A. Horn 1990 (see the References). These inequalities are the only results that are not given proofs in this book.

Problems

1. Use Theorem 7.9 to show that $A \geq B \geq 0$ implies

   $$\operatorname{rank}(A) \geq \operatorname{rank}(B), \quad \det A \geq \det B.$$

2. Let $A \geq 0$, $B \geq 0$. Prove or disprove $\lambda_i(A) \geq \lambda_i(B) \Rightarrow A \geq B$.

3. Show that $A > B \Rightarrow \lambda_i(A) > \lambda_i(B)$ for each i.

4. Let $A \geq B \geq 0$. Show that for $X \geq 0$ of the same size as A and B

   $$\lambda_i(AX) \geq \lambda_i(BX), \quad \text{for each } i.$$

5. Let $A \in \mathbb{M}_n$ be a positive semidefinite matrix. Show that

   $$\lambda_1(A)I - A \geq 0 \geq \lambda_n(A)I - A.$$

6. Let A and B be $n \times n$ positive semidefinite matrices. Show that

   $$\lambda_n(A) + \lambda_n(B) \leq \lambda_n(A+B) \leq \lambda_1(A+B) \leq \lambda_1(A) + \lambda_1(B);$$
   $$\lambda_n(A)\lambda_n(B) \leq \lambda_n(AB) \leq \lambda_1(AB) \leq \lambda_1(A)\lambda_1(B);$$
   $$\lambda_n(A)\lambda_n(B) \leq \lambda_n(A \circ B) \leq \lambda_1(A \circ B) \leq \lambda_1(A)\lambda_1(B).$$

7. Show by example that for $A \geq 0$ and $B \geq 0$ of the same size the following inequalities do not hold in general:
$$\lambda_i(A)\lambda_n(B) \leq \lambda_i(A \circ B) \leq \lambda_i(A)\lambda_1(B).$$

8. Let $A, B \in \mathbb{M}_n$ be Hermitian matrices. Prove or disprove
$$\lambda_i(A+B) \leq \lambda_i(A) + \lambda_i(B).$$

9. Let $A, B \in \mathbb{M}_n$ be Hermitian matrices. Show that for any $\alpha \in [0,1]$
$$\lambda_{\min}(\alpha A + (1-\alpha)B) \geq \alpha\lambda_{\min}(A) + (1-\alpha)\lambda_{\min}(B)$$
and
$$\lambda_{\max}(\alpha A + (1-\alpha)B) \leq \alpha\lambda_{\max}(A) + (1-\alpha)\lambda_{\max}(B).$$

10. Let A be Hermitian and $B \geq 0$ of the same size. Show that
$$\lambda_{\min}(B)\lambda_i(A^2) \leq \lambda_i(ABA) \leq \lambda_{\max}(B)\lambda_i(A^2).$$

11. Let $A \in \mathbb{M}_n$ be Hermitian. Show that for any $u \in \mathbb{C}^n$
$$\lambda_{i+2}(A + u^*u) \leq \lambda_{i+1}(A) \leq \lambda_i(A + u^*u)$$
and
$$\lambda_{i+2}(A) \leq \lambda_{i+1}(A + u^*u) \leq \lambda_i(A).$$

12. Let A be a Hermitian matrix partitioned as
$$A = \begin{pmatrix} A_{11} & A_{12} \\ A_{21} & A_{22} \end{pmatrix},$$
where A_{11} is a square submatrix. Show that
$$\begin{pmatrix} A_{11} & 0 \\ 0 & A_{22} \end{pmatrix} = \frac{1}{2}(A + UAU^*), \text{ where } U = \begin{pmatrix} I & 0 \\ 0 & -I \end{pmatrix},$$
and that
$$\lambda(A_{11} \oplus A_{22}) \prec \lambda(A).$$

13. When does $d(A) = \lambda(A)$? In other words, for what matrices are the diagonal entries equal to the eigenvalues? How about
$$d(A) = \sigma(A) \quad \text{or} \quad \lambda(A) = \sigma(A)?$$

14. Let A be an $n \times n$ positive semidefinite matrix. If V is an $n \times m$ matrix so that $V^*V = \operatorname{diag}(\delta_1, \delta_2, \ldots, \delta_m)$, each $\delta_i > 0$, show that
$$\lambda_n(A)\min_i \delta_i \leq \lambda_i(V^*AV) \leq \lambda_1(A)\max_i \delta_i.$$

15. Let A and B be n-square complex matrices. Show that
$$\sigma_n(A)\sigma_n(B) \leq |\lambda(AB)| \leq \sigma_1(A)\sigma_1(B)$$
for any eigenvalue $\lambda(AB)$ of AB. In particular, $|\lambda(A)| \leq \sigma_{\max}(A)$.

16. Let A be an $n \times n$ Hermitian matrix. Show that
$$(x^*Ax)^2 \leq (x^*AA^*x)(x^*x), \quad x \in \mathbb{C}^n,$$
and that
$$\lambda_i\left(\frac{A+A^*}{2}\right) \leq (\lambda_i(AA^*))^{\frac{1}{2}}.$$
Does it follow that
$$\sigma_i\left(\frac{A^*+A}{2}\right) \leq \sigma_i(A)?$$

17. Let A and B be $n \times n$ matrices. Show the following: For $r+s \leq n-1$,

    (a) if A and B are Hermitian, then
    $$\lambda_{r+s+1}(A+B) \leq \lambda_{r+1}(A) + \lambda_{s+1}(B)$$
    and
    $$\lambda_{n-r-s}(A+B) \geq \lambda_{n-r}(A) + \lambda_{n-s}(B);$$

    (b) if A and B are positive semidefinite, then
    $$\lambda_{r+s+1}(AB) \leq \lambda_{r+1}(A)\lambda_{s+1}(B)$$
    and
    $$\lambda_{n-r-s}(AB) \geq \lambda_{n-r}(A)\lambda_{n-s}(B);$$

    (c) for singular values
    $$\sigma_{r+s+1}(A+B) \leq \sigma_{r+1}(A) + \sigma_{s+1}(B),$$
    $$\sigma_{r+s+1}(AB) \leq \sigma_{r+1}(A)\sigma_{s+1}(B),$$
    and
    $$\sigma_{n-r-s}(AB) \geq \sigma_{n-r}(A)\sigma_{n-s}(B).$$
    But it is false that
    $$\sigma_{n-r-s}(A+B) \geq \sigma_{n-r}(A) + \sigma_{n-s}(B).$$

7.5 A Triangle Inequality for the Matrix $(A^*A)^{\frac{1}{2}}$

This section studies the positive semidefinite matrix $(A^*A)^{\frac{1}{2}}$, denoted by $|A|$, where A is any complex matrix. The main result, due to R. C. Thompson, is that for any square matrices A and B of the same size, there exist unitary matrices U and V such that

$$|A+B| \leq U^*|A|U + V^*|B|V.$$

Our first observation is on the converse of the statement

$$A \geq B \quad \Rightarrow \quad \lambda_i(A) \geq \lambda_i(B).$$

The inequalities $\lambda_i(A) \geq \lambda_i(B)$ cannot ensure $A \geq B$. For example,

$$A = \begin{pmatrix} 3 & 0 \\ 0 & 2 \end{pmatrix}, \quad B = \begin{pmatrix} 1 & 0 \\ 0 & 3 \end{pmatrix}.$$

Then $\lambda_1(A) = 3 \geq \lambda_1(B) = 3$ and $\lambda_2(A) = 2 \geq \lambda_2(B) = 1$. But $A - B$, having a negative eigenvalue -1, is obviously not positive semidefinite. We have, however, the following result.

Theorem 7.17 *Let $A, B \in \mathbb{M}_n$ be Hermitian matrices. If*

$$\lambda_i(A) \geq \lambda_i(B)$$

for each i, then there exists a unitary matrix U such that

$$U^*AU \geq B.$$

PROOF. Let P and Q be unitary matrices such that

$$A = P^* \operatorname{diag}(\lambda_1(A), \ldots, \lambda_n(A)) P,$$

$$B = Q^* \operatorname{diag}(\lambda_1(B), \ldots, \lambda_n(B)) Q.$$

The condition $\lambda_i(A) \geq \lambda_i(B)$, $i = 1, \ldots, n$, implies that

$$\operatorname{diag}(\lambda_1(A), \ldots, \lambda_n(A)) - \operatorname{diag}(\lambda_1(B), \ldots, \lambda_n(B)) \geq 0.$$

Multiply both sides by Q^* from the left and Q from the right to get

$$Q^* \operatorname{diag}(\lambda_1(A), \ldots, \lambda_n(A))Q - B \geq 0.$$

Take $U = P^*Q$. Then we have $U^*AU - B \geq 0$, as desired. ∎

We now turn our attention to the positive semidefinite matrix $|A|$. Note that $|A|$ is the unique positive semidefinite matrix satisfying

$$|A|^2 = A^*A.$$

Moreover, for any complex matrix A,

$$A = U|A| \tag{7.17}$$

is a polar decomposition of A, where U is some unitary matrix. Because of such a relation between $|A|$ and A, matrix $|A|$ has caught much attention, and many interesting results have been obtained.

Note that the eigenvalues of $|A|$ are the square roots of the eigenvalues of A^*A, namely, the singular values of A. In symbols,

$$\lambda(|A|) = \left(\lambda(A^*A)\right)^{\frac{1}{2}} = \sigma(A).$$

In addition, if $A = UDV$ is a singular value decomposition of A, then

$$|A| = V^*DV \quad \text{and} \quad |A^*| = UDU^*.$$

To prove Thompson's matrix triangle inequality, two theorems are needed. They are of interest in their own right.

Theorem 7.18 *Let A be an n-square complex matrix. Then*

$$\lambda_i\left(\frac{A^* + A}{2}\right) \leq \lambda_i(|A|), \quad i = 1, \ldots, n.$$

PROOF. Take $v_1, \ldots, v_n$ and $w_1, \ldots, w_n$ to be orthonormal sets of eigenvectors of $\frac{A^*+A}{2}$ and A^*A, respectively. Then for each i

$$\left(\frac{A^* + A}{2}\right)v_i = \lambda_i\left(\frac{A^* + A}{2}\right)v_i, \quad (A^*A)w_i = \left(\lambda_i(|A|)\right)^2 w_i.$$

For each fixed positive integer k with $1 \leq k \leq n$, let

$$S_1 = \text{Span}\{v_1, \ldots, v_k\}, \quad S_2 = \text{Span}\{w_k, \ldots, w_n\}.$$

Then for some unit vector $x \in S_1 \cap S_2$, by Theorem 7.5,

$$x^* \left(\frac{A^* + A}{2}\right) x \geq \lambda_k \left(\frac{A^* + A}{2}\right)$$

and

$$x^*(A^*A)x \leq \left(\lambda_k(|A|)\right)^2.$$

By using the Cauchy-Schwarz inequality it follows that

$$\begin{aligned}
\lambda_k\left(\frac{A^* + A}{2}\right) &\leq x^*\left(\frac{A^* + A}{2}\right)x \\
&= \text{Re}(x^*Ax) \\
&\leq |x^*Ax| \\
&\leq \sqrt{x^*A^*Ax} \\
&\leq \lambda_k(|A|). \quad \blacksquare
\end{aligned}$$

Combining Theorems 7.17 and 7.18, we see that for any n-square complex matrix A there exists a unitary matrix U such that

$$\frac{A^* + A}{2} \leq U^*|A|U. \tag{7.18}$$

We are now ready to give the matrix triangle inequality.

Theorem 7.19 (Thompson) *For any square complex matrices A and B of the same size, unitary matrices U and V exist such that*

$$|A + B| \leq U^*|A|U + V^*|B|V.$$

PROOF. By the polar decomposition (7.17), we may write

$$A + B = W|A + B|,$$

where W is a unitary matrix. Then for some unitary U, V by (7.18),

$$\begin{aligned} |A+B| &= W^*(A+B) \\ &= \frac{1}{2}\Big(W^*(A+B) + (A+B)^*W\Big) \\ &= \frac{1}{2}(A^*W + W^*A) + \frac{1}{2}(B^*W + W^*B) \\ &\leq U^*|W^*A|U + V^*|W^*B|V \\ &= U^*|A|U + V^*|B|V. \quad \blacksquare \end{aligned}$$

Note that the result would be false without the presence of the unitary matrices U and V (Problem 14).

Problems

1. Show that $\frac{A^*+A}{2}$ is Hermitian for any $n \times n$ matrix A and that

$$\frac{A^*+A}{2} \geq 0 \quad \Leftrightarrow \quad \operatorname{Re}(x^*Ax) \geq 0 \quad \text{for all } x \in \mathbb{C}^n.$$

2. Show that $A^*A \geq 0$ for any matrix A. What is the rank of $(A^*A)^{\frac{1}{2}}$?

3. Let A be a square complex matrix and let $|A| = (A^*A)^{\frac{1}{2}}$. Show that
    (a) A is positive semidefinite if and only if $|A| = A$,
    (b) A is normal if and only if $|A^*| = |A|$,
    (c) $|A|$ and $|A^*|$ are similar,
    (d) if $A = PU$ is a polar decomposition of A, where $P \geq 0$ and U is unitary, then $|A| = U^*PU$ and $|A^*| = P$.

4. Find $|A|$ and $|A^*|$ for each of the following matrices A:

$$\begin{pmatrix} 0 & 1 \\ 0 & 0 \end{pmatrix}, \quad \begin{pmatrix} 1 & 1 \\ 0 & 0 \end{pmatrix}, \quad \begin{pmatrix} 1 & 1 \\ 1 & 1 \end{pmatrix}, \quad \begin{pmatrix} 1 & 1 \\ -1 & -1 \end{pmatrix}.$$

5. Find $|A|$ and $|A^*|$ for $A = \begin{pmatrix} 1 & a \\ 0 & 0 \end{pmatrix}$, where $a > 0$.

6. Let $M = \begin{pmatrix} 0 & A^* \\ A & 0 \end{pmatrix}$. Show that $|M| = \begin{pmatrix} |A| & 0 \\ 0 & |A^*| \end{pmatrix}$.

7. Find $|A|$ for the normal matrix $A \in \mathbb{M}_n$ written as, with U unitary,
$$A = U\operatorname{diag}(\lambda_1,\ldots,\lambda_n)U^* = \sum_{i=1}^{n} \lambda_i u_i u_i^*.$$

8. Let A be an $m \times n$ complex matrix. Show that $|UAV| = V^*|A|V$ for any unitary matrices $U \in \mathbb{M}_m$ and $V \in \mathbb{M}_n$.

9. Let $A \in \mathbb{M}_n$. Then A and U^*AU have the same eigenvalues for any unitary $U \in \mathbb{M}_n$. Show that A and UAV have the same singular values for any unitary $U, V \in \mathbb{M}_n$. Do they have the same eigenvalues?

10. Let A be a Hermitian matrix. If X is a Hermitian matrix commuting with A such that $A \leq X$ and $-A \leq X$, show that $|A| \leq X$.

11. Show that for any matrix A there exist matrices X and Y such that
$$A = (AA^*)X \quad \text{and} \quad A = (AA^*)^{\frac{1}{2}}Y.$$

12. Let A be an $m \times n$ complex matrix, $m \geq n$. Show that there exists a unitary matrix U such that
$$AA^* = U^* \begin{pmatrix} A^*A & 0 \\ 0 & 0 \end{pmatrix} U.$$

13. Show that for any complex matrices A and B
$$|A \otimes B| = |A| \otimes |B|.$$

14. Construct an example showing that for $A, B \in \mathbb{M}_n$
$$|A + B| \leq |A| + |B|$$
is not true. Show that the trace inequality, however, holds:
$$\operatorname{tr}|A + B| \leq \operatorname{tr}(|A| + |B|).$$

15. Let A and B be n-square Hermitian matrices. Show that
$$\begin{pmatrix} A^2 & B^2 \\ B^2 & I \end{pmatrix} \geq 0 \quad \Rightarrow \quad \begin{pmatrix} |A| & |B| \\ |B| & I \end{pmatrix} \geq 0.$$

. ──────── ⊙ ──────── .

CHAPTER 8

Normal Matrices

INTRODUCTION A great deal of elegant work has been done for normal matrices. The goal of this chapter is to present basic results and methods on normal matrices. Section 8.1 gives conditions equivalent to the normality of matrices, Section 8.2 focuses on a special type of normal matrix with entries consisting of zeros and ones, Section 8.3 studies the positive semidefinite matrix $(A^*A)^{\frac{1}{2}}$ associated with a matrix A, and finally Section 8.4 shows majorization inequalities that, when equality holds, result in the normality of the matrix.

8.1 Equivalent Conditions

A square complex matrix A is said to be *normal* if it commutes with its conjugate transpose, in symbols,

$$A^*A = AA^*.$$

Matrix normality is one of the most interesting topics in linear algebra and matrix theory, since normal matrices have not only simple structures under unitary similarity but also many applications.

This section presents conditions equivalent to normality.

Theorem 8.1 Let $A = (a_{ij})$ be an n-square complex matrix with eigenvalues $\lambda_1, \lambda_2, \ldots, \lambda_n$. The following statements are equivalent:

1. A is normal, that is, $A^*A = AA^*$;

2. A is unitarily diagonalizable, namely, there exists a unitary matrix U of the same size such that
$$U^*AU = \mathrm{diag}(\lambda_1, \lambda_2, \ldots, \lambda_n); \tag{8.1}$$

3. there exists a polynomial $p(x)$ such that $A^* = p(A)$;

4. there exists a set of eigenvectors of A that form an orthonormal basis for $\mathbb{C}^n$;

5. every eigenvector of A is an eigenvector of A^*;

6. $A = B + iC$ for some B and C Hermitian, and $BC = CB$;

7. if U is a unitary matrix such that
$$U^*AU = \begin{pmatrix} B & C \\ 0 & D \end{pmatrix}$$
with B and D square, then B and D are normal and $C = 0$;

8. if $W \subseteq \mathbb{C}^n$ is an invariant subspace of A, then so is $W^\perp$;

9. if x is an eigenvector of A, then $x^\perp$ is invariant under A;

10. A can be written as
$$A = \sum_{i=1}^n \lambda_i E_i,$$
where $\lambda_i \in \mathbb{C}$ and $E_i \in \mathbb{M}_n$ satisfy
$$E_i^2 = E_i = E_i^*, \quad E_i E_j = 0 \ \text{if } i \neq j, \quad \sum_{i=1}^n E_i = I;$$

11. $\mathrm{tr}(A^*A) = \sum_{i,j=1}^n |a_{ij}|^2 = \sum_{i=1}^n |\lambda_i|^2$;

12. the singular values of A are $|\lambda_1|, |\lambda_2|, \ldots, |\lambda_n|$;

13. $\sum_{i=1}^{n} (\operatorname{Re} \lambda_i)^2 = \frac{1}{4} \operatorname{tr}(A + A^*)^2;$

14. $\sum_{i=1}^{n} (\operatorname{Im} \lambda_i)^2 = -\frac{1}{4} \operatorname{tr}(A - A^*)^2;$

15. the eigenvalues of $A + A^*$ are $\lambda_1 + \overline{\lambda_1}, \ldots, \lambda_n + \overline{\lambda_n};$

16. $\operatorname{tr}(A^*A)^2 = \operatorname{tr}\left((A^*)^2 A^2\right);$

17. $\|Ax\| = \|A^*x\|$ for every $x \in \mathbb{C}^n;$

18. $|A| = |A^*|$, where $|A| = (A^*A)^{\frac{1}{2}};$

19. $A^* = AU$ for some unitary $U;$

20. $A^* = VA$ for some unitary $V;$

21. $UP = PU$ if $A = UP$, a polar decomposition of $A;$

22. $AU = UA$ if $A = UP$, a polar decomposition of $A;$

23. $AP = PA$ if $A = UP$, a polar decomposition of $A;$

24. A commutes with $A + A^*;$

25. A commutes with $A - A^*;$

26. $A + A^*$ and $A - A^*$ commute;

27. A commutes with $A^*A;$

28. A commutes with $AA^* - A^*A;$

29. $A^*B = BA^*$ whenever $AB = BA;$

30. $A^*A - AA^*$ is positive semidefinite.

PROOF. $(2) \Leftrightarrow (1)$: We show that (1) implies (2). The other direction is obvious. Let $A = U^*TU$ be a Schur decomposition of A. It suffices to show that the upper-triangular matrix T is diagonal.

Note that $A^*A = AA^*$ yields $T^*T = TT^*$. By comparing the diagonal entries on both sides of the latter identity, we have $t_{ij} = 0$ whenever $i < j$. Thus T is diagonal.

SEC. 8.1   EQUIVALENT CONDITIONS   243

(3)⇔(2): To show that (2) implies (3), we choose a polynomial $p(x)$ of degree at most $n-1$, by interpolation, such that

$$p(\lambda_i) = \overline{\lambda_i}, \quad i = 1, \ldots, n.$$

Thus, if $A = U^* \operatorname{diag}(\lambda_1, \ldots, \lambda_n) U$ for some unitary matrix U, then

$$\begin{aligned}
A^* &= U^* \operatorname{diag}(\overline{\lambda_1}, \ldots, \overline{\lambda_n}) U \\
&= U^* \operatorname{diag}(p(\lambda_1), \ldots, p(\lambda_n)) U \\
&= U^* p(\operatorname{diag}(\lambda_1, \ldots, \lambda_n)) U \\
&= p(U^* \operatorname{diag}(\lambda_1, \ldots, \lambda_n) U) \\
&= p(A).
\end{aligned}$$

For the other direction, if $A^* = p(A)$ for some polynomial p, then

$$A^*A = p(A)A = Ap(A) = AA^*.$$

(4)⇔(2): If (8.1) holds, then multiplying by U from the left gives

$$AU = U \operatorname{diag}(\lambda_1, \ldots, \lambda_n)$$

or

$$Au_i = \lambda_i u_i, \quad i = 1, \ldots, n,$$

where u_i is the ith column of U, $i = 1, \ldots, n$. Thus, the column vectors of U are eigenvectors of A and they form an orthonormal basis of $\mathbb{C}^n$ since U is a unitary matrix.

Conversely, if A has a set of eigenvectors that form an orthonormal basis for $\mathbb{C}^n$, then the matrix U consisting of these vectors as columns is unitary and satisfies (8.1).

(5)⇔(1): Assume that A is normal and let u be an eigenvector of A corresponding to eigenvalue λ. Extend u to a unitary matrix with u as the first column; then

$$U^*AU = \begin{pmatrix} \lambda & \alpha \\ 0 & A_1 \end{pmatrix}. \tag{8.2}$$

The normality of A forces $\alpha = 0$.

Taking the conjugate transpose and by a simple computation, u is an eigenvector of A^* corresponding to the eigenvalue $\bar{\lambda}$ of A^*.

To see the other way around, we use induction on n. Note that

$$Ax = \lambda x \quad \Leftrightarrow \quad (U^*AU)(U^*x) = \lambda(U^*x)$$

for any n-square unitary matrix U. Thus, when considering $Ax = \lambda x$, we may assume that A is upper-triangular by Schur decomposition.

Take $e_1 = (1, 0, \ldots, 0)^T$. Then e_1 is an eigenvector of A. Hence, by assumption, e_1 is an eigenvector of A^*. A direct computation yields that the first column of A^* must consist of zeros except the first component. Thus, if we write

$$A = \begin{pmatrix} \lambda_1 & 0 \\ 0 & B \end{pmatrix}, \quad \text{then} \quad A^* = \begin{pmatrix} \bar{\lambda}_1 & 0 \\ 0 & B^* \end{pmatrix}.$$

Since every eigenvector of A is an eigenvector of A^*, this property is inherited by B and B^*. An induction hypothesis on B shows that A is diagonal. It follows that A is normal.

(6)$\Leftrightarrow$(1): It is sufficient to notice that $B = \frac{A+A^*}{2}$ and $C = \frac{A-A^*}{2i}$.

(7)$\Leftrightarrow$(1): We show that (1) implies (7). The other direction is easy. Upon computation, we have that $A^*A = AA^*$ implies

$$\begin{pmatrix} B^*B & B^*C \\ C^*B & C^*C + D^*D \end{pmatrix} = \begin{pmatrix} BB^* + CC^* & CD^* \\ DC^* & DD^* \end{pmatrix}.$$

Therefore,

$$B^*B = BB^* + CC^* \quad \text{and} \quad C^*C + D^*D = DD^*.$$

By taking the trace for both sides of the first identity and noticing that $\mathrm{tr}(BB^*) = \mathrm{tr}(B^*B)$, we obtain $\mathrm{tr}(CC^*) = 0$. This forces $C = 0$. Thus B is normal, and so is D by the second identity.

(8)$\Rightarrow$(9)$\Rightarrow$(4) and (7)$\Rightarrow$(8): It suffices to note that $\mathbb{C}^n = W \oplus W^\perp$ and that a basis of W and a basis of $W^\perp$ form a basis of $\mathbb{C}^n$.

(10)$\Rightarrow$(1): It is by a direct computation. To show (2)$\Rightarrow$(10), we write $U = (u_1, \ldots, u_n)$, where u_i is the ith column of U. Then

$$A = U \mathrm{diag}(\lambda_1, \ldots, \lambda_n) U^* = \lambda_1 u_1 u_1^* + \cdots + \lambda_n u_n u_n^*.$$

SEC. 8.1   EQUIVALENT CONDITIONS   245

Take $E_i = u_i u_i^*$, $i = 1, \ldots, n$. (10) then follows.

(11)$\Leftrightarrow$(2): Let $A = U^*TU$ be a Schur decomposition of A, where U is unitary and T is upper-triangular. Then $A^*A = U^*T^*TU$. Hence, $\operatorname{tr}(A^*A) = \operatorname{tr}(T^*T)$. On the other hand, upon computation,

$$\operatorname{tr}(A^*A) = \sum_{i,j=1}^{n} |a_{ij}|^2, \quad \operatorname{tr}(T^*T) = \sum_{i=1}^{n} |\lambda_i|^2 + \sum_{i<j} |t_{ij}|^2.$$

Thus, $\operatorname{tr}(A^*A) = \sum_{i=1}^{n} |\lambda_i|^2$ if and only if $t_{ij} = 0$ for all $i < j$, that is, T is diagonal and hence A is unitarily diagonalizable.

(12)$\Rightarrow$(11): If the singular values of A are $\sigma_1, \ldots, \sigma_n$, then

$$\begin{aligned}\operatorname{tr}(A^*A) &= \lambda_1(A^*A) + \cdots + \lambda_n(A^*A) \\ &= \sigma_1^2 + \cdots + \sigma_n^2 \\ &= |\lambda_1|^2 + \cdots + |\lambda_n|^2,\end{aligned}$$

which is (11). For the other direction, obviously (11)$\Rightarrow$(2)$\Rightarrow$(12).

(13)$\Rightarrow$(11): We may assume that A is an upper-triangular matrix, since the identity holds when A is replaced by U^*AU, where U is any unitary matrix. Notice that

$$\operatorname{tr}(A + A^*)^2 = \operatorname{tr} A^2 + 2\operatorname{tr}(A^*A) + \operatorname{tr}(A^*)^2.$$

It follows that

$$\operatorname{tr}(A^*A) = \frac{1}{2}\Big(\operatorname{tr}(A+A^*)^2 - \operatorname{tr} A^2 - \operatorname{tr}(A^*)^2\Big).$$

With this identity, (13) implies (11). (13) follows from (2).

Similarly, (14)$\Rightarrow$(11) and (2)$\Rightarrow$(14).

(15)$\Rightarrow$(13): If the eigenvalues of $A + A^*$ are $\lambda_1 + \overline{\lambda_1}, \ldots, \lambda_n + \overline{\lambda_n}$, then their squares are the eigenvalues of $(A + A^*)^2$. Thus,

$$\begin{aligned}\operatorname{tr}(A+A^*)^2 &= \sum_{i=1}^{n}(\lambda_i + \overline{\lambda_i})^2 \\ &= 4\sum_{i=1}^{n}(\operatorname{Re}\lambda_i)^2.\end{aligned}$$

(15) follows from (2) at once.

(16) is immediate from (1). To see the other way, we make use of the facts that for any matrices X and Y of the same size,
$$\operatorname{tr}(XY) = \operatorname{tr}(YX)$$
and
$$\operatorname{tr}(X^*X) = 0 \quad \Leftrightarrow \quad X = 0.$$
Upon computation, we have, noting that $\operatorname{tr}(AA^*)^2 = \operatorname{tr}(A^*A)^2$,
$$\operatorname{tr}\left((A^*A - AA^*)^*(A^*A - AA^*)\right) = \operatorname{tr}(A^*A - AA^*)^2$$
$$= \operatorname{tr}(A^*A)^2 - \operatorname{tr}\left((A^*)^2 A^2\right) - \operatorname{tr}\left(A^2(A^*)^2\right) + \operatorname{tr}(AA^*)^2,$$
which equals 0 by assumption. Thus, $A^*A - AA^* = 0$.

(17)$\Leftrightarrow$(1): The norm identity in (17) is rewritten, by squaring both sides, as the inner product identity
$$(Ax, Ax) = (A^*x, A^*x),$$
which is equivalent to
$$(x, A^*Ax) = (x, AA^*x),$$
or
$$\left(x, (A^*A - AA^*)x\right) = 0.$$
This holds for all $x \in \mathbb{C}^n$ if and only if $A^*A - AA^* = 0$.

(18)$\Leftrightarrow$(1): This is by the uniqueness of the square root.

(19)$\Leftrightarrow$(1): If $A^* = AU$ for some unitary U, then
$$A^*A = A^*(A^*)^* = (AU)(AU)^* = AA^*,$$
and A is normal. For the converse, we show (2)$\Rightarrow$(19). Let
$$A = V^* \operatorname{diag}(\lambda_1, \ldots, \lambda_n) V,$$
where V is unitary.

Take
$$U = V^* \operatorname{diag}(l_1, \ldots, l_n)V,$$
where $l_i = \frac{\overline{\lambda_i}}{\lambda_i}$ if $\lambda_i \neq 0$, and $l_i = 1$ otherwise, for $i = 1, \ldots, n$. Then
$$\begin{aligned} A^* &= V^* \operatorname{diag}(\overline{\lambda_1}, \ldots, \overline{\lambda_n})V \\ &= V^* \operatorname{diag}(\lambda_1, \ldots, \lambda_n)V \, V^* \operatorname{diag}(l_1, \ldots, l_n)V \\ &= AU. \end{aligned}$$

Similarly, (20) is equivalent to (1).

(21)$\Leftrightarrow$(1): If $A = UP$, where U is unitary and P is positive semidefinite, then $A^*A = AA^*$ implies
$$P^*P = UPP^*U^* \quad \text{or} \quad P^2 = UP^2U^*.$$
By taking square roots, we have
$$P = UPU^* \quad \text{or} \quad PU = UP.$$
The other direction is easy to check: $A^*A = P^2 = AA^*$.

(22)$\Leftrightarrow$(21): Note that U is invertible.

(23)$\Leftrightarrow$(21): We show that (23) implies (21). The other direction is immediate by multiplying P from the right-hand side.

Suppose $AP = PA$, that is, $UP^2 = PUP$. If P is nonsingular, then obviously $UP = PU$. Let $r = \operatorname{rank}(A) = \operatorname{rank}(P)$ and write
$$P = V^* \begin{pmatrix} D & 0 \\ 0 & 0 \end{pmatrix} V,$$
where V is unitary and D is $r \times r$ positive definite diagonal, $r < n$. Then $UP^2 = PUP$ gives, with $W = VUV^*$,
$$W \begin{pmatrix} D^2 & 0 \\ 0 & 0 \end{pmatrix} = \begin{pmatrix} D & 0 \\ 0 & 0 \end{pmatrix} W \begin{pmatrix} D & 0 \\ 0 & 0 \end{pmatrix}.$$
Partition W as, with $W_1 \in \mathbb{M}_r$,
$$\begin{pmatrix} W_1 & W_2 \\ W_3 & W_4 \end{pmatrix}.$$

Then
$$W_1 D^2 = D W_1 D \quad \text{and} \quad W_3 D^2 = 0,$$
which imply, for D is nonsingular,
$$W_1 D = D W_1 \quad \text{and} \quad W_3 = 0.$$
It follows that $W_2 = 0$ since W is unitary and that
$$W \begin{pmatrix} D & 0 \\ 0 & 0 \end{pmatrix} = \begin{pmatrix} D & 0 \\ 0 & 0 \end{pmatrix} W.$$
This results in $UP = PU$ at once.

(24, (25), and (26) are equivalent to (1) by direct computation.

(27)$\Rightarrow$(16): If A commutes with A^*A, then
$$AA^*A = A^*A^2.$$
Multiply both sides by A^* from the left to get
$$(A^*A)^2 = (A^*)^2 A^2.$$
Take the trace of both sides to obtain (16).

(28) is similarly proven.

(29)$\Leftrightarrow$(1): Take $B = A$ for the normality. To see the other direction, suppose that A is normal and that A and B commute, and let $A = U^* \operatorname{diag}(\lambda_1, \ldots, \lambda_n) U$ by (2). Then $AB = BA$ implies
$$\operatorname{diag}(\lambda_1, \ldots, \lambda_n)(UBU^*) = (UBU^*) \operatorname{diag}(\lambda_1, \ldots, \lambda_n).$$
Denote $T = UBU^* = (t_{ij})$. Then
$$\operatorname{diag}(\lambda_1, \ldots, \lambda_n) T = T \operatorname{diag}(\lambda_1, \ldots, \lambda_n),$$
which gives $(\lambda_i - \lambda_j) t_{ij} = 0$; thus, $(\overline{\lambda_i} - \overline{\lambda_j}) t_{ij} = 0$, for all i and j, which, in return, implies that $A^*B = BA^*$.

(30)$\Leftrightarrow$(1): This is a combination of two facts: $\operatorname{tr}(XY - YX) = 0$ for all square matrices X and Y of the same size; and if matrix X is positive semidefinite, then $\operatorname{tr} X = 0$ if and only if $X = 0$. ∎

Sec. 8.1 Equivalent Conditions 249

As many as 70 equivalent conditions of normal matrices have been observed in the literature. One will see more in the following exercises and in the later sections.

Problems

1. Show that each of the following conditions are equivalent to the normality of matrix $A \in \mathbb{M}_n$:

    (a) A has a linearly independent set of n eigenvectors, and any two corresponding to distinct eigenvalues are orthogonal;

    (b) every eigenvector of A is an eigenvector of $A + A^*$;

    (c) every eigenvector of A is an eigenvector of $A - A^*$;

    (d) $(Ax, Ay) = (A^*x, A^*y)$ for all $x, y \in \mathbb{C}^n$;

    (e) $(Ax, Ax) = (A^*x, A^*x)$ for all $x \in \mathbb{C}^n$.

2. Let A be an $n \times n$ complex matrix with eigenvalues λ_i and singular values σ_i arranged as $|\lambda_1| \geq \cdots \geq |\lambda_n|$ and $\sigma_1 \geq \cdots \geq \sigma_n$. Show that
$$\sigma_1 \sigma_2 \cdots \sigma_k = |\lambda_1 \lambda_2 \cdots \lambda_k|, \quad k = 1, 2, \ldots, n, \quad \Leftrightarrow \quad A^*A = AA^*.$$

3. Let A be a normal matrix. Show that $Ax = 0$ if and only if $A^*x = 0$.

4. Let A be a normal matrix. Show that if x is an eigenvector of A, then A^*x is also an eigenvector of A for the same eigenvalue.

5. If matrix A commutes with some normal matrix with distinct eigenvalues, show that A is normal. Is the converse true?

6. Show that unitary matrices, Hermitian matrices, skew-Hermitian matrices, real orthogonal matrices, and permutation matrices are all normal. Is a complex orthogonal matrix normal?

7. When is a normal matrix Hermitian? Positive semidefinite? Skew-Hermitian? Unitary? Nilpotent? Idempotent?

8. When is a triangular matrix normal?

9. Let A be a square matrix. Show that if A is a normal matrix, then $f(A)$ is normal for any polynomial f. If $f(A)$ is normal for some nonzero polynomial f, does it follow that A is normal?

10. Show that (27) is equivalent to (23) using Problem 27, Section 6.1.

11. Show that two normal matrices are similar if and only if they have the same set of eigenvalues and if and only if they are unitarily similar.

12. Let A and B be normal matrices of the same size. If $AB = BA$, show that AB is normal and that there exists a unitary matrix U that diagonalizes both A and B.

13. Let A be a nonsingular matrix and let $M = A^{-1}A^*$. Show that A is normal if and only if M is unitary.

14. Let $A \in \mathbb{M}_n$ be a normal matrix. Show that for any unitary $U \in \mathbb{M}_n$
$$\min_i\{|\lambda_i(A)|\} \leq |\lambda_i(AU)| \leq \max_i\{|\lambda_i(A)|\}.$$

15. Let A be a normal matrix. If $A^k = I$ for some positive integer k, show that A is unitary.

16. Let B be an n-square matrix and let A be the block matrix
$$A = \begin{pmatrix} B & B^* \\ B^* & B \end{pmatrix}.$$

Show that A is normal and that if B is normal with eigenvalues $\lambda_t = x_t + y_t i$, $x_t, y_t \in \mathbb{R}$, $t = 1, 2, \ldots, n$, then
$$\det A = 4i \prod_{t=1}^n x_t y_t.$$

17. Show that $\mathbb{C}^n = \text{Ker } A \oplus \text{Im } A$ for any n-square normal matrix A.

18. Let A and B be $n \times n$ normal matrices. If $\text{Im } A \perp \text{Im } B$, that is, $(x, y) = 0$ for all $x \in \text{Im } A$, $y \in \text{Im } B$, show that $A + B$ is normal.

19. Let A be Hermitian, B be skew-Hermitian, and $C = A + B$. Show that the following statements are equivalent:

    (a) C is normal; (b) $AB = BA$; (c) AB is skew-Hermitian.

20. Show that for any n-square complex matrix A
$$\text{tr}(A^*A)^2 \geq \text{tr}\left((A^*)^2 A^2\right).$$

Equality holds if and only if A is normal. Is it true that
$$(A^*A)^2 \geq (A^*)^2 A^2?$$

8.2 Normal Matrices with Zero and One Entries

This section presents two theorems on matrices with zero and one entries: One is on normality, and the other is on commutativity. Matrices of zeros and ones are referred to as *(0,1)-matrices*.

Throughout this section, J_n, or simply J, denotes the n-square matrix all of whose entries are 1. As usual, I is the identity matrix.

The following theorem often appears in combinatorics when a configuration of subsets is under investigation.

Theorem 8.2 *Let A be an n-square (0,1)-matrix. If*

$$AA^T = tI + J \qquad (8.3)$$

for some positive integer t, then A is a normal matrix.

PROOF. By considering the diagonal entries, we see that (8.3) implies that each row sum of A equals $t+1$, that is,

$$AJ = (t+1)J. \qquad (8.4)$$

Matrix A is nonsingular, since the determinant

$$(\det A)^2 = \det(AA^T) = \det(tI + J) = (t+n)t^{n-1}$$

is nonzero. Thus by (8.4), we have

$$A^{-1}J = (t+1)^{-1}J.$$

Multiplying both sides of (8.3) by J from the right gives

$$AA^T J = tJ + J^2 = (t+n)J.$$

It follows by multiplying A^{-1} from the left that

$$A^T J = (t+1)^{-1}(t+n)J.$$

By taking the transpose, we have

$$JA = (t+1)^{-1}(t+n)J. \qquad (8.5)$$

Multiply both sides by J from the right to get

$$JAJ = n(t+1)^{-1}(t+n)J.$$

Multiply both sides of (8.4) by J from the left to get

$$JAJ = n(t+1)J.$$

It follows by comparison of the right-hand sides that

$$(t+1)^2 = t+n \quad \text{or} \quad n = t^2 + t + 1.$$

Substituting this into (8.5), one has

$$JA = (t+1)J.$$

From (8.4),
$$AJ = JA \quad \text{or} \quad A^{-1}JA = J.$$

We then have

$$\begin{aligned} A^T A &= A^{-1}(AA^T)A \\ &= A^{-1}(tI + J)A \\ &= tI + A^{-1}JA \\ &= tI + J \\ &= AA^T. \quad \blacksquare \end{aligned}$$

Our next result asserts that if the product of two (0, 1)-matrices is a matrix whose diagonal entries are all zero and whose off-diagonal entries are all one, then these two matrices commute. The proof of this theorem uses the determinantal identity (Problem 5, Section 2.2)

$$\det(I_m + AB) = \det(I_n + BA),$$

where A and B are $m \times n$ and $n \times m$ matrices, respectively.

Theorem 8.3 *Let A and B be n-square (0,1)-matrices such that*

$$AB = J_n - I_n. \tag{8.6}$$

Then
$$AB = BA.$$

PROOF. Let a_i and b_j be the columns of A and B^T, respectively:

$$A = (a_1, \ldots, a_n), \quad B^T = (b_1, \ldots, b_n).$$

Then, by computation,

$$0 = \operatorname{tr}(AB) = \operatorname{tr}(BA) = \sum_{i=1}^n b_i^T a_i.$$

Thus $b_i^T a_i = 0$ for each i, since A and B have nonnegative entries.
Rewrite equation (8.6) as

$$I_n = J_n - AB \quad \text{or} \quad I_n = J_n - \sum_s a_s b_s^T.$$

Then

$$I_n + a_i b_i^T + a_j b_j^T = J_n - \sum_{s \neq i,\, j} a_s b_s^T.$$

Notice that the right-hand side contains $n-1$ matrices, each of which is of rank one. Thus, the matrix on the right-hand side has rank at most $n-1$, by the rank formula for sum (Problem 5). Hence the matrix on the left-hand side is singular. We have

$$\begin{aligned}
0 &= \det(I_n + a_i b_i^T + a_j b_j^T) \\
&= \det\left(I_n + (a_i, a_j)(b_i, b_j)^T\right) \\
&= \det\left(I_2 + (b_i, b_j)^T (a_i, a_j)\right) \\
&= \det\left(\begin{pmatrix} 1 & 0 \\ 0 & 1 \end{pmatrix} + \begin{pmatrix} 0 & b_i^T a_j \\ b_j^T a_i & 0 \end{pmatrix}\right) \\
&= 1 - (b_i^T a_j)(b_j^T a_i).
\end{aligned}$$

This forces $b_i^T a_j = 1$ for each pair of i and j, $i \neq j$, since A and B are $(0,1)$-matrices. It follows, by combining with $b_i^T a_i = 0$, that

$$BA = (b_i^T a_j) = J_n - I_n = AB. \blacksquare$$

Problems

1. Show that $A \in \mathbb{M}_n$ is normal if and only if $I - A$ is normal.

2. Let A be an n-square $(0, 1)$-matrix. Denote the number of 1 in row i by r_i and in column j by c_j. Show that

   A is normal $\Rightarrow$ $r_i = c_i$ for each i $\Leftrightarrow$ $J_n - A$ is normal.

3. Construct a nonnormal 4×4 normal matrix of zeros and ones such that each row and each column sum equal 2.

4. Let A be a 3×3 $(0, 1)$-matrix. Show that $\det A$ equals 0, ± 1, or ± 2.

5. Let $A_1, \ldots, A_n$ be $m \times m$ matrices each with rank 1. Show that

   $$\operatorname{rank}(A_1 + \cdots + A_n) \leq n.$$

6. Does there exist a normal matrix with real entries of the sign pattern

   $$\begin{pmatrix} + & + & 0 \\ + & 0 & + \\ + & + & + \end{pmatrix}?$$

7. If A is an $n \times n$ matrix with integer entries, show that $2Ax = x$ has no nontrivial solutions, that is, the only solution is $x = 0$.

8. Let A be a square $(0, 1)$-matrix such that $AA^T = tI + J$, and let $C = J - A$. Show that A commutes with C and C^T. Compute CC^T.

9. Let A be a square matrix such that $AA^T = tI + J$. Find the eigenvalues and singular values of A. When is A nonsingular?

10. Let A be a $v \times v$ matrix with zero and one entries such that

    $$AA^T = (k - h)I + hJ,$$

    where v, k, and h are positive integers satisfying $0 < h < k < v$.

    (a) Show that A is normal.
    (b) Show that $h = \frac{1}{v-1}k(k-1)$.
    (c) Show that $A^{-1} = \frac{1}{k(k-1)}(kA^T - hJ)$.
    (d) Find the eigenvalues of AA^T.

8.3 A Cauchy-Schwarz Type Inequality for Matrix $(A^*A)^{\frac{1}{2}}$

There is a variety of Cauchy-Schwarz inequalities of different types. The matrix version of the Cauchy-Schwarz inequality has been obtained using different methods and techniques. In this section we give some Cauchy-Schwarz matrix inequalities concerning the matrix

$$|A| = (A^*A)^{\frac{1}{2}}.$$

Obviously, if $A = UDV$ is a singular value decomposition of A, where U and V are unitary and D is nonnegative diagonal, then

$$|A| = V^*DV \quad \text{and} \quad |A^*| = UDU^*.$$

Theorem 8.4 *Let A be an n-square matrix. Then for any $u, v \in \mathbb{C}^n$,*

$$|(Au, v)|^2 \leq (|A|u, u)(|A^*|v, v) \tag{8.7}$$

and

$$|((A \circ A^*)u, u)| \leq ((|A| \circ |A^*|)u, u). \tag{8.8}$$

PROOF. It is sufficient, by Theorem 6.26, to observe that

$$\begin{pmatrix} |A| & A^* \\ A & |A^*| \end{pmatrix} = \begin{pmatrix} V^* & 0 \\ 0 & U \end{pmatrix} \begin{pmatrix} D & D \\ D & D \end{pmatrix} \begin{pmatrix} V & 0 \\ 0 & U^* \end{pmatrix} \geq 0.$$

and, by Theorem 6.21,

$$\begin{pmatrix} |A| \circ |A^*| & A^* \circ A \\ A \circ A^* & |A^*| \circ |A| \end{pmatrix} = \begin{pmatrix} |A| & A^* \\ A & |A^*| \end{pmatrix} \circ \begin{pmatrix} |A^*| & A \\ A^* & |A| \end{pmatrix} \geq 0.$$

Note that $|A| \circ |A^*| = |A^*| \circ |A|$, with $u = v$, gives (8.8). ∎

For more results, consider Hermitian matrices A decomposed as

$$A = U^* \operatorname{diag}(\lambda_1, \ldots, \lambda_n) U,$$

where U is unitary. Define A^α for $\alpha \in \mathbb{R}$, if each λ_i^α makes sense, as

$$A^\alpha = U^* \operatorname{diag}(\lambda_1^\alpha, \ldots, \lambda_n^\alpha) U.$$

Theorem 8.5 *Let $A \in \mathbb{M}_n$. Then for any $\alpha \in (0,1)$,*

$$|(Au, v)| \leq \||A|^\alpha u\| \, \||A^*|^{1-\alpha} v\|, \quad u, v \in \mathbb{C}^n. \tag{8.9}$$

PROOF. Let $A = U|A|$ be a polar decomposition of A. Then

$$A = U|A|^{1-\alpha} U^* U |A|^\alpha = |A^*|^{1-\alpha} U |A|^\alpha.$$

By the Cauchy-Schwarz inequality, we have

$$|(Au, v)| = |(U|A|^\alpha u, |A^*|^{1-\alpha} v)| \leq \||A|^\alpha u\| \, \||A^*|^{1-\alpha} v\|. \blacksquare$$

Note that (8.7) is a special case of (8.9) by taking $\alpha = \frac{1}{2}$.

Theorem 8.6 *Let $A \in \mathbb{M}_n$ and $\alpha \in \mathbb{R}$ be different from $\frac{1}{2}$. If*

$$|(Au, u)| \leq (|A|u, u)^\alpha (|A^*|u, u)^{1-\alpha}, \quad \text{for all } u \in \mathbb{C}^n, \tag{8.10}$$

then A is normal. The converse is obviously true.

PROOF. We first consider the case where A is nonsingular.

Let $A = UDV$ be a singular value decomposition of A, where D is diagonal and invertible, and U and V are unitary. With $|A| = V^* D V$ and $|A^*| = UDU^*$, the inequality in (8.10) becomes

$$|(UDVu, u)| \leq (V^* DVu, u)^\alpha (UDU^* u, u)^{1-\alpha}$$

or

$$|(D^{\frac{1}{2}} V u, D^{\frac{1}{2}} U^* u)| \leq (D^{\frac{1}{2}} V u, D^{\frac{1}{2}} V u)^\alpha (D^{\frac{1}{2}} U^* u, D^{\frac{1}{2}} U^* u)^{1-\alpha}.$$

For any nonzero $u \in \mathbb{C}^n$, set

$$y = \frac{1}{\|D^{\frac{1}{2}} U^* u\|} D^{\frac{1}{2}} U^* u.$$

Then $\|y\| = 1$, and y ranges over all unit vectors as u runs over all nonzero vectors. By putting $\hat{A} = D^{\frac{1}{2}} V U D^{-\frac{1}{2}}$, we have

$$|(\hat{A} y, y)| \leq (\hat{A} y, \hat{A} y)^\alpha, \quad \text{for all unit } y \in \mathbb{C}^n.$$

SEC. 8.3   A CAUCHY-SCHWARZ TYPE INEQUALITY FOR MATRIX $(A^*A)^{\frac{1}{2}}$

Applying Problem 19 of Section 5.1 to $\hat{A}$, we see $\hat{A}$ is unitary. Thus,

$$D^{-\frac{1}{2}}U^*V^*DVUD^{-\frac{1}{2}} = I.$$

It follows that
$$UDU^* = V^*DV.$$

By squaring both sides, we have $AA^* = A^*A$, or A is normal.

We next deal with the case where A is singular using mathematical induction on n. If $n = 1$, we have nothing to prove. Suppose that the assertion is true for $(n-1)$-square matrices.

Noting that (8.10) still holds when A is replaced by U^*AU for any unitary matrix U, we assume, without loss of generality, that

$$A = \begin{pmatrix} A_1 & b \\ 0 & 0 \end{pmatrix},$$

where $A_1 \in \mathbb{M}_{n-1}$ and b is an $(n-1)$-column vector.

If $b = 0$, then A is normal by induction on A_1. If $b \neq 0$, we take $u_1 \in \mathbb{C}^{n-1}$ such that $(b, u_1) \neq 0$. Let $u = \begin{pmatrix} u_1 \\ u_2 \end{pmatrix}$ with $u_2 > 0$. Then

$$|(Au, u)| = |(A_1 u_1, u_1) + (b, u_1)u_2|$$

and
$$(|A^*|u, u) = \left((A_1 A_1^* + bb^*)^{\frac{1}{2}} u_1, u_1\right)$$

which is independent of u_2. To compute $(|A|u, u)$, we write

$$|A| = \begin{pmatrix} C & d \\ d^* & \beta \end{pmatrix}.$$

Then
$$b^*b = d^*d + \beta^2.$$

Hence, $\beta \neq 0$; otherwise $d = 0$ and thus $b = 0$. Therefore, $\beta > 0$ and

$$(|A|u, u) = (Cu_1, u_1) + u_2\left((d, u_1) + (u_1, d)\right) + \beta u_2^2.$$

Letting $u_2 \to \infty$ in (8.10) implies $2\alpha \geq 1$ or $\alpha \geq \frac{1}{2}$.

(8.10) can be rewritten as, with A replaced by A^* and α by $1-\alpha$,

$$|(A^*u, u)| \leq (|A^*|u, u)^{1-\alpha}(|A|u, u)^{1-(1-\alpha)}.$$

Applying the same argument to A^*, one obtains $\alpha \leq \frac{1}{2}$. By the induction hypothesis, we see that A is normal if $\alpha \neq \frac{1}{2}$. ∎

Problems

1. Let A be a square complex matrix. Show that

   $$\operatorname{tr}|A| = \operatorname{tr}|A^*| \geq |\operatorname{tr} A| \quad \text{and} \quad \det|A| = \det|A^*| = |\det A|.$$

2. Give an example showing that the unitary matrix U is the decomposition $A = U^* \operatorname{diag}(\lambda_1, \ldots, \lambda_n) U$ is not unique. Show that the definition $A^\alpha = U^* \operatorname{diag}(\lambda_1^\alpha, \ldots, \lambda_n^\alpha) U$ for the Hermitian A and real α is independent of choices of unitary matrices U.

3. Show that for any n-square complex matrix A

   $$\begin{pmatrix} |A| & A^* \\ A & |A^*| \end{pmatrix} \geq 0, \quad \text{but} \quad \begin{pmatrix} |A| & A \\ A^* & |A^*| \end{pmatrix} \not\geq 0$$

   in general. Conclude that

   $$\begin{pmatrix} A & B \\ B^* & C \end{pmatrix} \geq 0 \quad \not\Rightarrow \quad \begin{pmatrix} A & B^* \\ B & C \end{pmatrix} \geq 0.$$

   Show that it is always true that

   $$\begin{pmatrix} A & B \\ B^* & A \end{pmatrix} \geq 0 \quad \Rightarrow \quad \begin{pmatrix} A & B^* \\ B & A \end{pmatrix} \geq 0.$$

4. For any n-square complex matrices A and B, show that

   $$\begin{pmatrix} |A|+|B| & A^*+B^* \\ A+B & |A^*|+|B^*| \end{pmatrix} \geq 0.$$

   Derive the determinantal inequality

   $$(\det|A+B|)^2 \leq \det(|A|+|B|)\det(|A^*|+|B^*|).$$

   In particular,
   $$\det|A+A^*| \leq \det(|A|+|A^*|).$$

   Discuss the analog for the Hadamard product.

SEC. 8.3   A CAUCHY-SCHWARZ TYPE INEQUALITY FOR MATRIX $(A^*A)^{\frac{1}{2}}$ 259

5. Let A be an n-square complex matrix and $\alpha \in [0,1]$. Show that
$$\begin{pmatrix} |A|^{2\alpha} & A^* \\ A & |A^*|^{2(1-\alpha)} \end{pmatrix} \geq 0.$$

6. Let A and B be n-square complex matrices. Show that
$$A^*A = B^*B \quad \Leftrightarrow \quad |A| = |B|.$$

Is it true that
$$A^*A \geq B^*B \quad \Leftrightarrow \quad |A| \geq |B|?$$

Prove or disprove
$$|A| \geq |B| \quad \Leftrightarrow \quad |A^*| \geq |B^*|.$$

7. Let $[A]$ denote a principal submatrix of a square matrix A. Show by example that $|[A]|$ and $[|A|]$ are not comparable, that is,
$$|[A]| \not\geq [|A|] \quad \text{and} \quad [|A|] \not\geq |[A]|.$$

But the inequalities below hold, assuming the inverses involved exist:

(a) $[|A|^2] \geq [|A|]^2$,
(b) $[|A|^2] \geq |[A]|^2$,
(c) $[|A|^{\frac{1}{2}}] \leq [|A|]^{\frac{1}{2}}$,
(d) $[|A|^{-\frac{1}{2}}] \geq [|A|]^{-\frac{1}{2}}$,
(e) $[|A|^{-1}] \leq [|A|^{-2}]^{\frac{1}{2}}$.

8. Show that an n-square complex matrix A is normal if and only if
$$|(Au, u)| \leq (|A|u, u), \quad u \in \mathbb{C}^n.$$

(Hint: Suppose A is upper-triangular and take $u = (1, 0, \ldots, 0)^T$.)

9. Let A, B, C, $D \in \mathbb{M}_n$. If $CC^* + DD^* \leq I_n$, show that
$$|CA + DB| \leq (|A|^2 + |B|^2)^{\frac{1}{2}}.$$

8.4 Majorization and Matrix Normality

We consider in this section the inequalities involving diagonal entries, eigenvalues, and singular values of matrices. The equality cases of these inequalities will result in the normality of the matrices.

Theorem 8.7 (Schur Inequality) *Let $A = (a_{ij})$ be an n-square complex matrix with eigenvalues $\lambda_1, \lambda_2, \ldots, \lambda_n$. Then*

$$\sum_{i=1}^n |\lambda_i|^2 \leq \sum_{i,j=1}^n |a_{ij}|^2.$$

Equality occurs if and only if A is normal.

PROOF. Let $A = U^*TU$ be a Schur decomposition of A, where U is unitary and T is upper-triangular. Then $A^*A = U^*T^*TU$; consequently, $\mathrm{tr}(A^*A) = \mathrm{tr}(T^*T)$. Upon computation, we have

$$\mathrm{tr}(A^*A) = \sum_{i,j=1}^n |a_{ij}|^2$$

and

$$\mathrm{tr}(T^*T) = \sum_{i=1}^n |\lambda_i|^2 + \sum_{i<j} |t_{ij}|^2.$$

The inequality is immediate. For the equality case, notice that each $t_{ij} = 0$, $i < j$, that is, T is diagonal. Hence A is unitarily diagonalizable, thus normal. The other direction is obvious. ∎

An interesting application of this result is to show that if matrices A, B, and AB are normal, then so is BA (Problem 9).

Theorem 8.8 *Let $A = (a_{ij}) \in \mathbb{M}_n$ have singular values σ_i. Then*

$$|\mathrm{tr}\, A| \leq \sigma_1 + \cdots + \sigma_n. \tag{8.11}$$

Equality holds if and only if $A = uP$ for some $P \geq 0$ and some $u \in \mathbb{C}$ with $|u| = 1$; consequently, A is normal (but not conversely).

PROOF. Let $A = UDV$ be a singular value decomposition of A, where $D = \mathrm{diag}(\sigma_1, \ldots, \sigma_n)$ with $\sigma_1 \geq \cdots \geq \sigma_r > \sigma_{r+1} = 0 = \cdots = 0$, $r = \mathrm{rank}\,(A)$, and U and V are unitary. We then have, by computation,

$$\begin{aligned} |\operatorname{tr} A| &= \left| \sum_{i=1}^{n} \sum_{j=1}^{n} u_{ij} \sigma_j v_{ji} \right| \\ &= \left| \sum_{j=1}^{n} \sum_{i=1}^{n} u_{ij} v_{ji} \sigma_j \right| \\ &\leq \sum_{j=1}^{n} \left| \sum_{i=1}^{n} u_{ij} v_{ji} \right| \sigma_j \\ &\leq \sum_{j=1}^{n} \left(\sum_{i=1}^{n} |u_{ij} v_{ji}| \right) \sigma_j \\ &\leq \sum_{j=1}^{n} \sigma_j. \end{aligned}$$

The last inequality was due to the Cauchy-Schwarz inequality:

$$\sum_{i=1}^{n} |u_{ij} v_{ji}| \leq \sqrt{\sum_{i=1}^{n} |u_{ij}|^2 \sum_{i=1}^{n} |v_{ji}|^2} = 1.$$

If equality holds for the overall inequality, then

$$\left| \sum_{i=1}^{n} u_{ij} v_{ji} \right| = \sum_{i=1}^{n} |u_{ij} v_{ji}| = 1, \quad \text{for each } j \leq r.$$

Rewrite $\sum_{i=1}^{n} u_{ij} v_{ji}$ as (u_j, v_j^*), where u_j is the jth column of U and v_j is the jth row of V. By the equality case of the Cauchy-Schwarz inequality, it follows that $u_j = c_j v_j^*$, for each $j \leq r$, where c_j is a constant with $|c_j| = 1$. Thus $|\operatorname{tr} A|$ equals, by switching v_i and v_i^*,

$$|\operatorname{tr}(c_1 \sigma_1 v_1^* v_1 + \cdots + c_r \sigma_r v_r^* v_r)| = |c_1 \sigma_1 v_1 v_1^* + \cdots + c_r \sigma_r v_r v_r^*|.$$

Notice that $v_i v_i^* = 1$ for each i. By Problem 3, we have $A = c_1 V^* DV$. The other direction is easy to verify. ∎

We now turn our attention to the majorization inequalities of diagonal entries, eigenvalues, and singular values.

Recall that for real vectors $x = (x_1, \ldots, x_n)$ and $y = (y_1, \ldots, y_n)$ with components in decreasing order, x is weakly majorized by y, in symbols, $x \prec_w y$, if the partial sum of the first k largest components of x is dominated by that of y for $k = 1, \ldots, n$, and x is majorized by y, written $x \prec y$, if, in addition, equality holds for $k = n$.

For complex $x = (x_1, \ldots, x_n)$, assume $|x_1| \geq \cdots \geq |x_n|$ and write $|x| = (|x_1|, \ldots, |x_n|)$. For a square complex matrix A, we denote by $d(A)$, $\lambda(A)$, and $\sigma(A)$, as before, the vectors of diagonal entries, eigenvalues, and singular values of A, respectively.

Theorem 8.9 *Let A be an n-square complex matrix. Then*

$$|d(A)| \prec_w \sigma(A) \tag{8.12}$$

and

$$|\lambda(A)| \prec_w \sigma(A). \tag{8.13}$$

Moreover, the first $\prec_w$ becomes $\prec$ if and only if

$$A = PU$$

for some positive semidefinite matrix P and some diagonal unitary matrix U, and the second $\prec_w$ becomes $\prec$ if and only if A is normal.

PROOF. We may assume that the diagonal entries of A are in decreasing order of absolute value. For each a_{ii}, let t_i be such that

$$t_i a_{ii} = |a_{ii}|, \quad |t_i| = 1, \quad i = 1, \ldots, n.$$

Let $B = A \operatorname{diag}(t_1, \ldots, t_n)$ and let C be the leading $k \times k$ principal submatrix of B. Then B has the same singular values as A, and

$$d(C) = (|a_{11}|, \ldots, |a_{kk}|).$$

By Theorem 7.12, we have $\sigma_i(C) \leq \sigma_i(B)$, $i = 1, \ldots, k$.

Applying the preceding theorem, we have

$$\begin{aligned} |\operatorname{tr} C| &= |a_{11}| + \cdots + |a_{kk}| \\ &\leq \sigma_1(C) + \cdots + \sigma_k(C) \\ &\leq \sigma_1(B) + \cdots + \sigma_k(B) \\ &= \sigma_1(A) + \cdots + \sigma_k(A). \end{aligned}$$

Thus, $\prec_w$ in (8.12) follows. If $\prec$ occurs, then

$$\sum_{i=1}^n |a_{ii}| = \sum_{i=1}^n \sigma_i(A)$$

and

$$|\operatorname{tr} B| = \sum_{i=1}^n \sigma_i(A) = \sum_{i=1}^n \sigma_i(B).$$

It follows from Theorem 8.8 that $B = uP$ for some positive semidefinite matrix P and some constant u with $|u| = 1$. Therefore,

$$A = B \operatorname{diag}(\overline{t_1}, \ldots, \overline{t_n}) = PU, \quad \text{where } U = u \operatorname{diag}(\overline{t_1}, \ldots, \overline{t_n}).$$

Conversely, if $A = PU$ with $P \geq 0$ and U diagonal unitary, then one has $|d(A)| = d(P)$. Since $\sigma(A) = \sigma(P) = \lambda(P)$, $\prec_w$ is in fact $\prec$.

For (8.13), let $A = U^*TU$ be a Schur decomposition of A, where U is unitary and T is upper-triangular. Then

$$|\lambda(A)| = |d(T)| \quad \text{and} \quad \sigma(A) = \sigma(T).$$

An application of (8.12) gives the $\prec_w$ in (8.13). If $\prec$ occurs, then T is necessarily diagonal, and A is normal. The converse is easy. ∎

Problems

1. Let $A \in \mathbb{M}_n$. Show the stronger inequalities than that in (8.11):

$$|\operatorname{tr} A| \leq \sum_{i=1}^n |a_{ii}| \leq \sigma_1 + \cdots + \sigma_n.$$

2. Let A be n-square complex matrix. Show that for each k, $1 \leq k \leq n$,

$$\sum_{i=1}^k \sum_{j=1}^n |a_{ij}|^2 \leq \sum_{i=1}^k \sigma_i^2(A).$$

3. Let $p_1, \ldots, p_n$ be positive numbers and $c_1, \ldots, c_n$ be complex. If

$$|c_1 p_1 + \cdots + c_n p_n| = |c_1| p_1 + \cdots + |c_n| p_n,$$

show that there exists $\theta \in \mathbb{R}$ such that $e^{i\theta} c_k \geq 0$ for all $k = 1, \ldots, n$.

4. Let A and B be n- and m-square complex matrices, respectively, and let P be an $m \times n$ matrix such that $P^*P = I_n$. Show that for each i
$$\sigma_i(PA) = \sigma_i(A) \quad \text{and} \quad \sigma_i(BP) \leq \sigma_i(B).$$

5. Let A be an n-square matrix with singular values $\sigma_i(A)$. Show that
$$|\det A|^{\frac{1}{n}} \leq \frac{1}{n} \sum_{i=1}^{n} \sigma_i(A)$$
with equality if and only if A is normal.

6. If some entry a_{ij} of a matrix A is the largest singular value of A, show that row i and column j contain entirely zero except a_{ij}.

7. Let $A \in \mathbb{M}_n$ have eigenvalues λ_i and singular values σ_i. Show that
$$\sum_{i=1}^{n} |\lambda_i| = \sum_{i=1}^{n} \sigma_i \Leftrightarrow |\lambda_i| = \sigma_i \text{ for each } i \Leftrightarrow A \text{ is normal}.$$

8. Let A be an n-square complex matrix. Show that
$$\lambda_i(A^2) = \left(\lambda_i(A)\right)^2, \quad \text{for each } i,$$
and
$$\sum_{i=1}^{n} |\lambda_i(A)|^2 \leq \sum_{i=1}^{n} \sigma_i(A^2).$$
Is it true that
$$\sigma_i(A^2) = \left(\sigma_i(A)\right)^2, \quad \text{for each } i?$$

9. Let A and B be normal matrices of the same size. Show that
$$AB \text{ is normal} \quad \Rightarrow \quad BA \text{ is normal}.$$

10. (**Fan and Hoffman**) Let A be an $n \times n$ complex matrix. Show that
$$\lambda\left(\frac{A^* + A}{2}\right) \prec_w \left|\lambda\left(\frac{A^* + A}{2}\right)\right| \prec_w \sigma(A).$$

. ——————— ⊙ ——————— .

References

Books:

T. W. Anderson, *The Statistical Analysis of Time Series,* Wiley, New York, Reprint edition, 1994.

T. Ando, *Operator-Theoretic Methods for Matrix Inequalities,* Lecture Notes, Preprint, 1989.

R. B. Bapat and T. E. S. Raghavan, *Nonnegative Matrices and Applications,* Cambridge University Press, New York, 1997.

E. F. Beckenbach and R. Bellman, *Inequalities,* Springer-Verlag, New York, Fourth printing, 1983.

R. Bellman, *Introduction to Matrix Analysis,* SIAM, Philadelphia, Reprint of the Second edition, 1995.

A. Berman and R. Plemmons, *Nonnegative Matrices in the Mathematical Sciences,* SIAM, Philadelphia, 1994.

R. Bhatia, *Matrix Analysis,* Springer-Verlag, New York, 1997.

R. Brualdi and H. J. Ryser, *Combinatorial Matrix Theory,* Cambridge University Press, New York, 1991.

D. Carlson, C. Johnson, D. Lay, A. Porter, A. Watkins and W. Watkins, *Resources for Teaching Linear Algebra,* Mathematical Association of America, Washington, DC , 1997.

P. J. Davis, *Circulant Matrices,* John Wiley & Sons, Inc., New York, 1979.

W. F. Donoghue Jr., *Monotone Matrix Functions and Analytic Continuation,* Springer-Verlag, New York, 1974.

F. R. Gantmacher, *The Theory of Matrices, Volume One,* Chelsea Publishing Company, New York, 1964.

S. K. Godunov, *Modern Aspects of Linear Algebra,* Translations of Mathematical Monographs, Vol. 175, American Mathematical Society, Providence, RI, 1998.

I. Gohberg, P. Lancaster and L. Rodman, *Matrix Polynomials,* Academic Press, New York, 1982.

G. H. Golub and C. F. Van Loan, *Matrix Computations,* Johns Hopkins University Press, Baltimore, 3rd edition, 1996.

F. Graybill, *Matrices with Applications in Statistics,* Wadsworth, Belmont, California, 1983.

P. R. Halmos, *Linear Algebra Problem Book,* Mathematical Association of America, Washington, DC , 1995.

D. A. Harville, *Matrix Algebra from a Statistician's Perspective,* Springer-Verlag, New York, 1997.

N. J. Higham, *Handbook of Writing for the Mathematical Sciences,* SIAM, Philadelphia, 1993.

References

R. A. Horn and C. R. Johnson, *Matrix Analysis,* Cambridge University Press, New York, 1985.

R. A. Horn and C. R. Johnson, *Topics in Matrix Analysis,* Cambridge University Press, New York, 1991.

M. S. Klamkin, *Problems in Applied Mathematics,* SIAM, Philadelphia, 1990.

A. I. Kostrikin, *Exercises in Algebra,* Gordon and Breach Publishers, Amsterdam, 1996.

P. Lancaster and M. Tismenetsky, *The Theory of Matrices,* Second edition, Academic Press, Orlando, 1985.

P. D. Lax, *Linear Algebra,* John Wiley & Sons, Inc., New York, 1997.

J.-S. Li and J.-G. Cha, *Linear Algebra (in Chinese),* University of Science and Technology of China Press, Hefei, China, 1989.

S.-Z. Liu, *Contributions to Matrix Calculus and Applications in Econometrics,* Tinbergen Institute Research Series, no. 106, Thesis Publishers, Amsterdam, 1995.

J. Magnus and H. Neudecker, *Matrix Differential Calculus with Applications in Statistics and Econometrics,* John Wiley & Sons, Inc., New York, 1988.

M. Marcus and H. Minc, *A Survey of Matrix Theory and Matrix Inequalities,* Reprint edition, Dover, New York, 1992.

A. W. Marshall and I. Olkin, *Inequalities: Theory of Majorization and Its Applications,* Academic Press, New York, 1979.

M. L. Mehta, *Matrix Theory,* Hindustan Publishing Company, New Delhi, India, 1989.

R. Merris, *Multilinear Algebra,* Gordon & Breach Publishers, Amsterdam, 1997.

H. Minc, *Permanents,* Addison-Wesley, New York, 1978.

L. Mirsky, *An Introduction to Linear Algebra,* Reprint edition, Dover, New York, 1990.

S. Montgomery and E. W. Ralston, *Selected Papers on Algebra, Volume Three,* Mathematical Association of America, Washington, DC, 1977.

G.-X. Ni, *Common Methods in Matrix Theory (in Chinese),* Shanghai Science and Technology Press, Shanghai, 1984.

V. V. Prasolov, *Problems and Theorems in Linear Algebra,* American Mathematical Society, Providence, RI, 1994.

C. R. Rao and M. B. Rao, *Matrix Algebra and Its Applications to Statistics and Econometrics,* World Scientific Publishing, Singapore, 1998.

W. Rudin, *Principles of Mathematical Analysis,* McGraw-Hill, New York, 1976.

J. R. Schott, *Matrix Analysis for Statistics,* John Wiley & Sons, Inc., New York, 1997.

M.-R. Shi, *600 Linear Algebra Problems with Solutions (in Chinese),* Beijing Press of Science and Technology, Beijing, 1985.

G. W. Stewart and J.-G. Sun, *Matrix Permutation Theory,* Academic Press, New York, 1990.

G. Strang, *Linear Algebra and Applications,* Academic Press, New York, 1976.

F. Uhlig and R. Grone, *Current Trends in Matrix Theory,* North-Holland, New York, 1986.

B.-Y. Wang, *Introduction to Majorization Inequalities (in Chinese)*, Beijing Normal University Press, Beijing, 1991.

S.-G. Wang and Z.-Z. Jia, *Inequalities in Matrix Theory (in Chinese)*, Anhui Education Press, Hefei, China, 1994.

S.-F. Xu, *Theory and Methods in Matrix Computation (in Chinese)*, Beijing University Press, Beijing, 1995.

Y.-C. Xu, *Introduction to Algebra (in Chinese)*, Shanghai Science and Technology Press, Shanghai, 1982.

H. Yanai, *Projection Matrices, Generalized Inverses, and Singular Value Decompositions*, University of Tokyo Press, Tokyo, Second printing, 1993.

F.-Z. Zhang, *Linear Algebra: Challenging Problems for Students*, Johns Hopkins University Press, Baltimore, MD, 1996.

Papers:

C. Akemann, J. Anderson and G. Pedersen, *Triangle inequalities in operator algebras*, Linear and Multilinear Algebra, Vol. 11, pp. 167–178, 1982.

G. Alpargu and G. P. H. Styan, *Some comments and a future bibliography on the Frucht-Kantorovich and Wielandt inequalities, and on some related inequalities*, Preprint, 1998.

T. W. Anderson and G. P. H. Styan, *Cochran's theorem, rank additivity and tripotent matrices*, Statistics and Probability: Essays in Honor of C. R. Rao, Edited by G. Kallianpur, P. R. Krishnaiah and J. K. Ghosh, North-Holland Publishing Company, pp. 1–23, 1982.

T. Ando, *Concavity of certain maps on positive definite matrices and applications to Hadamard products*, Linear Algebra and Its Applications, Vol. 26, pp. 203–241, 1979.

T. Ando, *Hölder type inequalities for matrices*, Mathematical Inequalities and Applications, Vol. 1, no. 1, pp. 1–30, 1998.

T. Ando, R. A. Horn and C. R. Johnson, *The singular values of a Hadamard product: A basic inequality*, Linear and Multilinear Algebra, Vol. 21, pp. 345–365, 1987.

Y. H. Au-Yeung and Y. T. Poon, 3×3 *orthostochastic matrices and the convexity of generalized numerical ranges*, Linear Algebra and Its Applications, Vol. 27, pp. 69–79, 1979.

C. S. Ballantine, *A note on the matrix equation* $H = AP + PA^*$, Linear Algebra and Its Applications, Vol. 2, pp. 37–47, 1969.

R. B. Bapat and M. K. Kwong, *A generalization of* $A \circ A^\mathsf{T} \geq I$, Linear Algebra and Its Applications, Vol. 93, pp. 107–112, 1987.

R. B. Bapat and V. S. Sunder, *On majorization and Schur products*, Linear Algebra and Its Applications, Vol. 72, pp. 107–117, 1985.

S. J. Bernau and G. G. Gregory, *A Cauchy-Schwarz inequality for determinants*, American Mathematical Monthly, Vol. 84, pp. 495–496, June–July 1977.

R. Bhatia, R. Horn and F. Kittaneh, *Normal approximations to binormal operators*, Linear Algebra and Its Applications, Vol. 147, pp. 169–179, 1991.

R. Brualdi and H. Schneider, *Determinantal identities: Gauss, Schur, Cauchy, Sylvester, Kronecker, Jacobi, Binet, Laplace, Muir, and Cayley*, Linear Algebra and Its Applications, Vol. 52/53, pp. 769–791, 1983.

D. Carlson, C. R. Johnson, D. Lay and A. D. Porter, *Gems of exposition in elementary linear algebra*, College Mathematical Journal, Vol. 23, no. 4, pp. 299–303, 1992.

N. N. Chan and M. K. Kwong, *Hermitian matrix inequalities and a conjecture*, American Mathematical Monthly, Vol. 92, pp. 533–541, October 1985.

R. Chapman, *A polynomial taking integer values*, Mathematics Magazine, Vol. 69, no. 2, p. 121, April 1996.

M.-D. Choi, *A Schwarz inequality for positive linear maps on C^*-algebras*, Illinois Journal of Mathematics, Vol. 18, pp. 565–574, 1974.

J. Chollet, *Some Inequalities for principal submatrices*, American Mathematical Monthly, Vol. 104, pp. 609–617, August–September 1997.

C. Davis, *Notions generalizing convexity for functions defined on spaces of matrices*, Proceedings of Symposia in Pure Mathematics, American Mathematical Society, Vol. 7, pp. 187–201, 1963.

D. Z. Djokovic, *On some representations of matrices*, Linear and Multilinear Algebra, Vol. 4, pp. 33–40, 1979.

S. W. Drury, *A bound for the determinant of certain Hadamard products and for the determinant of the sum of two normal matrices*, Linear Algebra and Its Applications, Vol. 199, pp. 329–338, 1994.

C. F. Dunkl and K. S. Williams, *A simple norm inequality*, American Mathematical Monthly, Vol 71, pp. 53–54, January 1964.

L. Elsner, *The generalized spectral-radius theorem: An analytic-geometric proof*, Linear Algebra and Its Applications, Vol. 220, pp. 151–159, 1995.

K. Fan, *On a theorem of Weyl concerning eigenvalues of linear transformations I*, Proceedings of the National Academy of Sciences U.S.A., Vol. 35, pp. 652–655, 1949.

M. Fiedler, *A note on the Hadamard product of matrices*, Linear Algebra and Its Applications, Vol. 49, pp. 233–235, 1983.

S. Fisk, *A note on Weyl's inequality*, American Mathematical Monthly, Vol. 104, pp. 257–258, March 1997.

T. Furuta, $A \geq B \geq 0$ *assures* $(B^r A^p B^r)^{1/q} \geq B^{(p+2r)/q}$ *for* $r \geq 0, p \geq 0, q \geq 1$ *with* $(1+2r)q \geq p+2r$, Proceedings of the American Mathematical Society, Vol. 101, pp. 85–88, September 1987.

G. R. Goodson, *The inverse-similarity problem for real orthogonal matrices*, American Mathematical Monthly, Vol. 104, pp. 223–230, March 1997.

W. Govaerts and J. D. Pryce, *A singular value inequality for block matrices*, Linear Algebra and Its Applications, Vol. 125, pp. 141–148, 1989.

R. Grone, C. R. Johnson, E. M. Sa and H. Wolkowicz, *Normal matrices*, Linear Algebra and Its Applications, Vol. 87, pp. 213–225, 1987.

R. Grone, S. Pierce and W. Watkins., *Extremal correlation matrices*, Linear Algebra and Its Applications, Vol. 134, pp. 63–70, 1990.

J. Groß, G. Trenkler and S. Troschke, *On a characterization associated with the matrix arithmetic and geometric means,* Image: Bulletin of the International Linear Algebra Society, no. 17, p. 32, Summer 1996.

P. R. Halmos, *Bad products of good matrices,* Linear and Multilinear Algebra, Vol. 29, pp. 1–20, 1991.

E. V. Haynsworth, *Applications of an inequality for the Schur complement,* Proceedings of the American Mathematical Society, Vol. 24, pp. 512–516, 1970.

Y. P. Hong and R. A. Horn, *A canonical form for matrices under consimilarity,* Linear Algebra and Its Applications, Vol. 102, pp. 143–168, 1988.

R. A. Horn, *The Hadamard product,* Proceedings of Symposia in Applied Mathematics, Vol. 40, Edited by C. R. Johnson, American Mathematical Society, Providence, RI, pp. 87–169, 1990.

R. A. Horn and I. Olkin, *When does $A^*A = B^*B$ and why does one want to know?* American Mathematical Monthly, Vol. 103, pp. 270–482, June–July 1996.

L.-K. Hua, *Inequalities involving determinants (in Chinese),* Acta Math Sinica, Vol. 5, no. 4, pp. 463–470, 1955.

Y. Ikebe, T. Inagaki, and S. Miyamoto, *The monotonicity theorem, Cauchy's interlace theorem, and the Courant-Fischer theorem,* American Mathematical Monthly, Vol. 94, pp. 352–354, April 1987.

E. I. Im, *Narrower eigenbounds for Hadamard products,* Linear Algebra and Its Applications, Vol. 264, pp. 141–144, 1997.

E.-X. Jiang, *Bounds for the smallest singular value of a Jordan block with an application to eigenvalue permutation,* Linear Algebra and Its Applications, Vol. 197/198, pp. 691–707, 1994.

C. R. Johnson, *An inequality for matrices whose symmetric part is positive definite,* Linear Algebra and Its Applications, Vol. 6, pp. 13–18, 1973.

C. R. Johnson, *Inverse M-matrices,* Linear Algebra and Its Applications, Vol. 47, pp. 195–216, 1982.

C. R. Johnson, *The relationship between AB and BA,* American Mathematical Monthly, Vol. 103, pp. 578–582, August–September 1996.

C. R. Johnson and F.-Z. Zhang, *An operator inequality and matrix normality,* Linear Algebra and Its Applications, Vol. 240, pp. 105–110, 1996.

K. R. Laberteaux, M. Marcus, G. P. Shannon, E. A. Herman, R. B. Israel and G. Letac, *Hermitian matrices,* Problem solutions, American Mathematical Monthly, Vol. 104, p. 277, March 1997.

C.-K. Li, *Matrices with some extremal properties,* Linear Algebra and Its Applications, Vol. 101, pp. 255–267, 1988.

R.-C. Li, *Norms of certain matrices with applications to variations of the spectra of matrices and matrix pencils,* Linear Algebra and Its Applications, Vol. 182, pp. 199–234, 1993.

Z.-S. Li, F. Hall and F.-Z. Zhang, *Sign patterns of nonnegative normal matrices,* Linear Algebra and Its Applications, Vol. 254, pp. 335–354, 1997.

S.-Z. Liu and H. Neudecker, *Several matrix Kantorovich-type inequalities,* Journal of Mathematical Analysis and Applications, Vol. 197, pp. 23–26, 1996.

D. London, *A determinantal inequality and its permanental counterpart*, Linear and Multilinear Algebra, Vol. 42, pp. 281–290, 1997.

M. Lundquist and W. Barrett, *Rank inequalities for positive semidefinite matrices*, Linear Algebra and Its Applications, Vol. 248, pp. 91–100, 1996.

M. Marcus, K. Kidman and M. Sandy, *Products of elementary doubly stochastic matrices*, Linear and Multilinear Algebra, Vol. 15, pp. 331–340, 1984.

T. Markham, *Oppenheim's inequality for positive definite matrices*, American Mathematical Monthly, Vol. 93, pp. 642–644, October 1986.

G. Marsaglia and G. P. H. Styan, *Equalities and inequalities for ranks of matrices*, Linear and Multilinear Algebra, Vol. 2, pp. 269–292, 1974.

A. W. Marshall and I. Olkin, *Reversal of the Lyapunov, Hölder, and Minkowski inequalities and other extensions of the Kantorovich inequality*, Journal of Mathematical Analysis and Applications, Vol. 8, pp. 503–514, 1964.

A. W. Marshall and I. Olkin, *Matrix versions of the Cauchy and Kantorovich inequalities*, Aequationes Mathematicae, Vol. 40, pp. 89–93, 1990.

R. Mathias, *Concavity of monotone matrix functions of finite order*, Linear and Multilinear Algebra, Vol. 27, pp. 129–138, 1990.

R. Mathias, *An arithmetic-geometric-harmonic mean inequality involving Hadamard products*, Linear Algebra and Its Applications, Vol. 184, pp. 71–78, 1993.

C. McCarthy and A. T. Benjamin, *Determinants of tournaments*, Mathematics Magazine, Vol. 69, no. 2, pp. 133–135, April 1996.

J. K. Merikoski, *On the trace and the sum of elements of a matrix*, Linear Algebra and Its Applications, Vol. 60, pp. 177–185, 1984.

D. Merino, *Matrix similarity*, Image: Bulletin of the International Linear Algebra Society, no. 21, p. 26, October 1998.

R. Merris, *The permanental dominance conjecture*, in Current Trends in Matrix Theory, Edited by F. Uhlig and R. Grone, Elsevier, New York, pp. 213–223, 1987.

L. Mirsky, *A note on normal matrices*, American Mathematical Monthly, Vol. 63, p. 479, 1956.

B. Mond and J. E. Pečarić, *Matrix versions of some means inequalities*, Australian Mathematical Society, Vol. 20, pp. 117–120, 1993.

M. Newman, *On a problem of H. J. Ryser*, Linear and Multilinear Algebra, Vol. 12, pp. 291–293, 1983.

P. Nylen, T. Y. Tam and F. Uhlig, *On the eigenvalues of principal submatrices of normal, Hermitian and symmetric matrices*, Linear and Multilinear Algebra, Vol. 36, pp. 69–78, 1993.

K. Okubo, *Hölder-type norm inequalities for Schur products of matrices*, Linear Algebra and Its Applications, Vol. 91, pp. 13–28, 1987.

I. Olkin, *Symmetrized product definiteness?*, Image: Bulletin of the International Linear Algebra Society, no. 19, p. 32, Summer 1997.

D. V. Ouellette, *Schur complements and statistics*, Linear Algebra and Its Applications, Vol. 36, pp. 187–295, 1981.

C. C. Paige, G. P. H. Styan, B.-Y. Wang and F.-Z. Zhang, *Revisiting Hua's matrix equality and related inequalities, Schur complements and Sylvester's law of inertia*, Preprint, 1998.

W. V. Parker, *Sets of complex numbers associated with a matrix*, Duke Mathematical Journal, Vol. 15, pp. 711–715, 1948.

R. Patel and M. Toda, *Trace inequalities involving Hermitian matrices*, Linear Algebra and Its Applications, Vol. 23, pp. 13–20, 1979.

J. E. Pečarić, S. Puntanen and G. P. H. Styan, *Some further matrix extensions of the Cauchy-Schwarz and Kantorovich inequalities, with some statistical applications*, Linear Algebra and Its Applications, Vol. 237/238, pp. 455–476, 1996.

V. Pták, *The Kantorovich inequality*, American Mathematical Monthly, Vol. 102, pp. 820–821, November 1995.

S. M. Robinson, *A short proof of Cramer's rule*, Mathematics Magazine, Vol. 43, pp. 94–95, 1970.

L. Rodman, *Products of symmetric and skew-symmetric matrices*, Linear and Multilinear Algebra, Vol. 43, pp. 19–34, 1997.

B. L. Shader and C. L. Shader, *Scheduling conflict-free parties for a dating service*, American Mathematical Monthly, Vol. 104, pp. 99–106, February 1997.

C. Shafroth, *A generalization of the formula for computing the inverse of a matrix*, American Mathematical Monthly, Vol. 88, pp. 614–616, October 1981.

W. So and R. C. Thompson, *Products of exponentials of Hermitian and complex symmetric matrices*, Linear and Multilinear Algebra, Vol. 29, pp. 225–233, 1991.

G. W. Soules, *An approach to the permanental-dominance conjecture*, Linear Algebra and Its Applications, Vol. 201, pp. 211–229, 1994.

J. Stewart, *Positive definite functions and generalizations: A historical survey*, Rocky Mountain J. of Mathematics, Vol. 6, pp. 409–434, 1976.

G. P. H. Styan, *Hadamard products and multivariate statistical analysis*, Linear Algebra and Its Applications, Vol. 6, pp. 217–240, 1973.

G. P. H. Styan, *Schur complements and linear statistical models*, Proceedings of the First International Tampere Seminar on Linear Statistical Models and Their Applications, Edited by T. Pukkila and S. Puntanen, Department of Mathematical Sciences, University of Tampere, Tampere, Finland, pp. 37–75, 1985.

J.-G. Sun, *On two functions of a matrix with positive definite Hermitian part*, Linear Algebra and Its Applications, Vol. 244, pp. 55–68, 1996.

R. C. Thompson, *Convex and concave functions of singular values of matrix sums*, Pacific Journal of Mathematics, Vol. 66, pp. 285–290, 1976.

R. C. Thompson, *Matrix type metric inequalities*, Linear and Multilinear Algebra, Vol. 5, pp. 303–319, 1978.

R. C. Thompson, *High, low, and quantitative roads in linear algebra*, Linear Algebra and Its Applications, Vol. 162–164, pp. 23–64, 1992.

G. Visick, *Majorizations of Hadamard products of matrix powers*, Linear Algebra and Its Applications, Vol. 269, pp. 233–240, 1998.

B.-Y. Wang and F.-Z. Zhang, *Words and matrix normality*, Linear and Multilinear Algebra, Vol. 34, pp. 93–89, 1995.

B.-Y. Wang and F.-Z. Zhang, *A trace inequality for unitary matrices,* American Mathematical Monthly, Vol. 101, no. 5, pp. 453–455, May 1994.

B.-Y. Wang and F.-Z. Zhang, *Trace and eigenvalue inequalities for ordinary and Hadamard products of positive semidefinite Hermitian matrices,* SIAM Journal on Matrix Theory and Applications, Vol. 16, pp. 1173–1183, October 1995.

B.-Y. Wang and F.-Z. Zhang, *Schur complements and matrix inequalities of Hadamard products,* Linear and Multilinear Algebra, Vol. 43, pp. 315–326, 1997.

W. Watkins, *A determinantal inequality for correlation matrices,* Linear Algebra and Its Applications, Vol. 79, pp. 209–213, 1988.

G. S. Watson, G. Alpargu, and G. P. H. Styan, *Some comments on six inequalities associated with the inefficiency of ordinary least squares with one regressor,* Linear Algebra and Its Applications, Vol. 264, pp. 13–53, 1997.

C.-S. Wong, *Characterizations of products of symmetric matrices,* Linear Algebra and Its Applications, Vol. 42, pp. 243–251, 1982.

P.-Y. Wu, *Products of positive semidefinite matrices,* Linear Algebra and Its Applications, Vol. 111, pp. 53–61, 1988.

R. Yarlagadda and J. Hershey, *A note on the eigenvectors of Hadamard matrices of order 2^n,* Linear Algebra and Its Applications, Vol. 45, pp. 43–53, 1982.

X.-Z. Zhan, *Inequalities for the singular values of Hadamard products,* SIAM Journal on Matrix Theory and Applications, Vol. 18, pp. 1093–1095, October 1997.

F.-Z. Zhang, *A determinantal inequality,* Image: Bulletin of the International Linear Algebra Society, no. 16, p. 32, Winter 1996.

F.-Z. Zhang, *Quaternions and matrices of quaternions,* Linear Algebra and Its Applications, Vol. 251, pp. 21–57, 1997.

F.-Z. Zhang, *Matrix similarity,* Image: Bulletin of the International Linear Algebra Society, no. 20, p. 32, Winter 1998.

Notation

$\mathbb{M}_n$	n-square complex matrices
$\mathbb{F}$	a field
$\mathbb{C}$	complex numbers
$\mathbb{R}$	real numbers
$\mathbb{Q}$	rational numbers
$\mathbb{C}^n$	column vectors with n complex components
$\mathbb{R}^n$	column vectors with n real components
$\mathbb{F}[x]$	polynomials over field $\mathbb{F}$
$\mathbb{F}_n[x]$	polynomials over field $\mathbb{F}$ with degree at most n
ω	nth primitive root of unity
$\operatorname{Re} c$	real part of complex number c
$\operatorname{Im} c$	imaginary part of complex number c
$\operatorname{Span} S$	vector space spanned by the vectors in S
$\dim V$	dimension of the vector space V
$V + W$	sum of subspaces V and W
$V \oplus W$	direct sum of subspaces V and W
e_i	vector with ith component 1 and 0 elsewhere
(u, v)	inner product of vectors u and v, i.e, v^*u
$\angle_{x,y}$	angle between vectors x and y
$\|x\|$	absolute value vector $\|x\| = (\|x_1\|, \ldots, \|x_n\|)$
$\\|x\\|$	length or norm of vector x, i.e., $\\|x\\| = \sqrt{x^*x} = (\sum_{i=1}^n \|x_i\|^2)^{\frac{1}{2}}$
$x^\perp$	vectors orthogonal to vector x
$S^\perp$	vector space orthogonal to set S
$S_1 \perp S_2$	$(x, y) = 0$ for all $x \in S_1$ and $y \in S_2$
V_λ	eigenspace of the eigenvalue λ
I_n, I	identity matrix of order n
$A = (a_{ij})$	matrix with entries a_{ij}
A^T	transpose of matrix A
$\overline{A}$	conjugate of matrix A
A^*	conjugate transpose of matrix A
A^{-1}	inverse of matrix A
A_{11}	principal submatrix of matrix A in the upper-left corner
$A(i\|j)$	matrix by deleting the ith row and jth column of matrix A
$\operatorname{adj}(A)$	adjoint matrix of matrix A
$\det A$	determinant of matrix A
$\operatorname{rank}(A)$	rank of matrix A
$\operatorname{tr} A$	trace of matrix A
$\operatorname{diag} S$	diagonal matrix with the elements of S on the diagonal
$\begin{vmatrix} A & B \\ C & D \end{vmatrix}$	determinant of the 2×2 block matrix

$(A, B)_{\mathrm{M}}$	matrix inner product, i.e., $(A, B)_{\mathrm{M}} = \mathrm{tr}(B^*A)$
Im A	image of A, i.e., Im $A = \{Ax\}$
Ker A	kernel or null space of A, i.e., Ker $A = \{x : Ax = 0\}$
$\mathcal{R}(A)$	space spanned by the row vectors of matrix A
$\mathcal{C}(A)$	space spanned by the column vectors of matrix A
$W(A)$	numerical range of matrix A
$w(A)$	numerical radius of matrix A
$\rho(A)$	spectral radius of matrix A
In(A)	inertia of Hermitian matrix A
T_n	n-square tridiagonal matrix
H_n	Hadamard matrix
$V_n(a_1, \ldots, a_n)$	n-square Vandermonde matrix of $a_1, \ldots, a_n$
$s_k(a_1, \ldots, a_n)$	kth symmetric function of $a_1, \ldots, a_n$
$\lambda_{\max}(A)$	largest eigenvalue of matrix A
$\sigma_{\max}(A)$	largest singular value of matrix A
$\lambda_{\min}(A)$	smallest eigenvalue of matrix A
$\sigma_{\min}(A)$	smallest singular value of matrix A
$\lambda_i(A)$	eigenvalue of matrix A
$\sigma_i(A)$	singular value of matrix A
$d(A)$	vector of diagonal entries of matrix A
$\lambda(A)$	eigenvalue vector of matrix A
$\sigma(A)$	singular value vector of matrix A
$d(\lambda)$	invariant factors of matrix A
$m_A(\lambda)$	minimal polynomial of matrix A
$\det(\lambda I - A)$	characteristic polynomial of matrix A
$p(\lambda) \mid q(\lambda)$	$p(\lambda)$ divides $q(\lambda)$
$A \geq 0$	A is positive semidefinite
$A > 0$	A is positive definite
$A \geq B$	$A - B$ is positive semidefinite
$A^{\frac{1}{2}}$	square root of positive semidefinite matrix A
A^α	$A^\alpha = U^* \mathrm{diag}(\lambda_1^\alpha, \ldots, \lambda_n^\alpha)U$ if $A = U^* \mathrm{diag}(\lambda_1, \ldots, \lambda_n)U$
$[A]$	principal submatrix of A
$\lvert A \rvert$	$\lvert A \rvert = (A^*A)^{\frac{1}{2}}$
$\widetilde{A_{11}}$	Schur complement of A_{11}
δ_{ij}	Kronecker numbers, i.e., $\delta_{ij} = 1$ if $i = j$, and 0 otherwise
$x \prec_w y$	x is weakly majorized by y
$x \prec y$	x is majorized by y
$x \circ y$	$x \circ y = (x_1 y_1, \ldots, x_n y_n)$
$A \oplus B$	direct sum of matrices A and B, i.e., $A \oplus B = \begin{pmatrix} A & 0 \\ 0 & B \end{pmatrix}$
$A \otimes B$	Kronecker product of matrices A and B
$A \circ B$	Hadamard product of matrices A and B
$\mathcal{A}, \ldots, \mathcal{Z}$	linear transformations
$P \Rightarrow Q$	statement P implies statement Q
$P \Leftrightarrow Q$	statements P and Q are equivalent

Index

addition 1
adjoint 10
angle 25
basis 2
Birkhoff G. 127
Cauchy interlacing theorem 221
Cauchy-Schwarz inequality 22, 203, 255
Cauchy sequence 142
Cayley-Hamilton 70
characteristic polynomial 18, 70
cofactor 10
column space 49
compact set 89, 138
complete space 142
conjugate 7
continuity argument 36, 56
contraction 142
convex combination 127
convex hull 92
convergence 142
coordinate 3
Courant-Fischer 220
decomposition
    Jordan 74, 78
    polar 67
    Schur 64
    singular value 66
    spectral 65
    triangular 65
determinant 9
differential operator 14
dimension 2
dimension identity 4
direct product 190
direct sum 5, 9
eigenspace 19
eigenvalue 16, 18, 51, 219, 227
eigenvector 16
elementary divisor 75

elementary operation 8, 30, 74
elementary symmetric function 113
Fan K. 231, 264
field of values 88
Fischer inequality 175
fixed point 143
Frobenius-König 126
Hadamard inequality 176
Hadamard product 190
Hoffman A. 264
Hua inequality 187
image 14
inertia 212
inner product 22
inner product space 22
interlacing theorem 222
interpolation 60
invariant factor 75
invariant subspace 19
inverse 10, 20, 43
involution 93
isomorphism 21
Jordan block 74
Jordan canonical form 74
Jordan form 74
Kantorovich inequality 204
kernel 14
Kronecker product 190
Laplace expansion 9
length 23
linear dependence 2
linear independence 2
linear transformation 14
Löwner partial ordering 166
LU factorization 68
majorization 229, 262
matrix 6, 15
    backward identity 13, 81
    block 7

matrix (continued)
  Cauchy 105, 197
  circulant 106
  complex orthogonal 131
  diagonal 7
  doubly stochastic 126
  elementary 8
  elementary λ- 75
  Fourier 108
  generalized elementary 30
  Hadamard 118
  Hermitian 7, 65, 208
  idempotent 93
  identity 7
  invertible 10
  invertible λ- 75
  involutary 93
  irreducible 123
  nilpotent 93
  nonnegative definite 160
  nonsingular 10
  normal 7, 65, 240
  orthogonal 7
  partitioned 7
  permutation 123
  positive definite 159
  positive semidefinite 65, 159
  primary permutation 106, 124
  projection 93
  real orthogonal 131
  reducible 123
  scalar 7
  skew-Hermitian 211
  skew-symmetric 57
  symmetric 7
  tridiagonal 101
  Topelitz 110
  unitary 7, 131
  upper-triangular 7
  Vandermonde 11, 110, 111
  zero 7
  (0, 1)- 251
  λ- 74
matrix addition 6
matrix product 6

metric 143
metric space 142
minimal polynomial 71
minor 20
multiplicity 51
norm 23
null space 14
numerical radius 89
numerical range 88
Oppenheim A. 200
orthogonal projection 95
orthogonal set 23
orthogonal vector 23
orthonormal set 23
permutation 130
permutation similarity 123
Poincaré interlacing theorem 221
primary permutation matrix 107, 124
primitive root 106
principal submatrix 20
projection 93, 95, 100
QR factorization 68
rank 9, 46
Rayleigh-Ritz 220
real orthogonal projection 140
reflection 137
rotation 137
row space 49
scalar multiplication 1, 6
Schur I. 192, 230
Schur complement 175, 184
Schur inequality 260
Schur product 190
similarity 15
singular value 55, 66, 90, 144, 228
solution space 9
span 2
spectral norm 90
spectral radius 90
square root 66, 144, 162
standard basis 3
standard form 75
strict contraction 142
Sturm interlacing theorem 221
submatrix 7

subspace 3
sum of subspaces 3
Sylvester J. 46
tensor product 190
Thompson R. C. 237
Toeplitz-Hausdorff 88
trace 18
transpose 7

triangle inequality 23
triangularization 65
unit vector 23
unitary similarity 64
vector 1
vector space 1
weak majorization 229, 262
Wielant inequality 207

Universitext *(continued)*

Ramsay/Richtmyer: Introduction to Hyperbolic Geometry
Reisel: Elementary Theory of Metric Spaces
Rickart: Natural Function Algebras
Rotman: Galois Theory
Rubel/Colliander: Entire and Meromorphic Functions
Sagan: Space-Filling Curves
Samelson: Notes on Lie Algebras
Schiff: Normal Families
Shapiro: Composition Operators and Classical Function Theory
Simonnet: Measures and Probability
Smith: Power Series From a Computational Point of View
Smoryński: Self-Reference and Modal Logic
Stillwell: Geometry of Surfaces
Stroock: An Introduction to the Theory of Large Deviations
Sunder: An Invitation to von Neumann Algebras
Tondeur: Foliations on Riemannian Manifolds
Wong: Weyl Transforms
Zhang: Matrix Theory: Basic Results and Techniques
Zong: Strange Phenomena in Convex and Discrete Geometry
Zong: Sphere Packings